INTERNATIONAL ENERGY AGENCY

RENEWABLE
SOURCES
OF
ENERGY

MARCH 1987

INTERNATIONAL ENERGY AGENCY

2, RUE ANDRÉ-PASCAL 75775 PARIS CEDEX 16, FRANCE

The International Energy Agency (IEA) is an autonomous body which was established in November 1974 within the framework of the Organisation for Economic Co-operation and Development (OECD) to implement an International Energy Program.

It carries out a comprehensive programme of energy co-operation among twenty-one* of the OECD's twenty-four Member countries. The basic aims of IEA are:

i)   co-operation among IEA Participating Countries to reduce excessive dependence on oil through energy conservation, development of alternative energy sources and energy research and development;

ii) an information system on the international oil market as well as consultation with oil companies;

iii) co-operation with oil producing and other oil consuming countries with a view to developing a stable international energy trade as well as the rational management and use of world energy resources in the interest of all countries;

iv) a plan to prepare Participating Countries against the risk of a major disruption of oil supplies and to share available oil in the event of an emergency.

*IEA Member countries: Australia, Austria, Belgium, Canada, Denmark, Germany, Greece, Ireland, Italy, Japan, Luxembourg, Netherlands, New Zealand, Norway, Portugal, Spain, Sweden, Switzerland, Turkey, United Kingdom, United States.

Pursuant to article 1 of the Convention signed in Paris on 14th December, 1960, and which came into force on 30th September, 1961, the Organisation for Economic Co-operation and Development (OECD) shall promote policies designed:

–   to achieve the highest sustainable economic growth and employment and a rising standard of living in Member countries, while maintaining financial stability, and thus to contribute to the development of the world economy;

–   to contribute to sound economic expansion in Member as well as non-member countries in the process of economic development; and

–   to contribute to the expansion of world trade on a multilateral, non-discriminatory basis in accordance with international obligations.

The original Member countries of the OECD are Austria, Belgium, Canada, Denmark, France, the Federal Republic of Germany, Greece, Iceland, Ireland, Italy, Luxembourg, the Netherlands, Norway, Portugal, Spain, Sweden, Switzerland, Turkey, the United Kingdom and the United States. The following countries became Members subsequently through accession at the dates indicated hereafter: Japan (28th April, 1964), Finland (28th January, 1969), Australia (7th June, 1971) and New Zealand (29th May, 1973).

The Socialist Federal Republic of Yugoslavia takes part in some of the work of the OECD (agreement of 28th October, 1961).

Table of Contents

ANNEX I

NATURE, STATUS AND OUTLOOK OF RENEWABLE ENERGY SOURCES AND CONVERSION TECHNOLOGIES

ANNEX II

ECONOMICS OF RENEWABLE ENERGY TECHNOLOGIES

FOREWORD

The concept of renewable sources of energy is attractive to IEA Member countries which are all pursuing the goal of redirecting their energy supplies away from imported oil through such activities as development of indigenous sources of energy. However, the large variation among countries in the availability of these sources and in their local economic situation causes each country's development plans to be unique. Some IEA members therefore are counting on large contributions from indigenous renewable energy sources; while others predict minimal effects from these sources.

This study reports the results of the first comprehensive analysis of renewable energy sources by the IEA. It attempts to understand what has been achieved by over ten years of concerted action by governments and industry and why the outlook for most renewable energy technologies is still very promising though long-term. This work should aid governments to adjust their expectations for the timing and size of each renewable energy source's contribution so that development priorities can be checked and realistic long-term efforts can be maintained.

The Secretariat was assisted in the preparation of this study by officials of Member governments and outside experts. I am most grateful to all of them for their help without which the study could not have been completed. The study is, however, published on my responsibility as Executive Director of the IEA and does not necessarily reflect the views or positions of the IEA or its Member governments.

Helga STEEG
Executive Director

RENEWABLE SOURCES OF ENERGY

Executive Summary

INTRODUCTION

This study reviews renewable energy sources in IEA Member countries and their related technologies in an attempt to assess:

— the accomplishments and lessons of over a decade of developments;

— the most promising areas of further RD&D and/or other government action to increase the development and utilisation of renewable energies; and

— the outlook for contributions to national energy supplies on an economic basis.

The renewable energy sources included in this study are solar, wind, biomass, geothermal and ocean energies. In general, the study covers all of the related technologies which are being or have been successfully developed for the conversion of renewable energy sources into usable forms of energy. Technologies which are generally considered mature and fully integrated with conventional energy sources, such as large hydropower, are not discussed.

The study describes the nature of each renewable energy source and covers, for each technology area: background on development; descriptions of sub-technologies and applications; technical status and accomplishments; economic status and commercialization; environmental considerations; and technology outlook. Two annexes provide the details of the technical and economic analyses upon which the main report's discussions are based. Institutional factors affecting market introduction and penetration and the types of government actions which have been effective in accelerating development of renewable energy technologies are discussed more generally. Finally, additional R&D efforts and opportunities for international collaboration on that R&D are identified, and findings, conclusions and recommendations are provided.

BACKGROUND

The use of renewable energy sources is, of course, not new, these being the first energy sources to be used by man. However, the movement systematically to explore and use these resources as a component of world energy supplies began about a decade ago, after the oil crises of the early 1970s. At that time, renewable energy technologies appeared attractive for industry because of the projected high cost of oil and because of their promise of modularity, rapid deployability and attendant financial risk reduction. For governments, the promotion of renewable energy technology development had several motivations. Many countries began seeking to improve their energy security through the use of indigenous resources such as renewable sources of energy to reduce their need for imported oil. Being widely distributed geographically and apparently inexhaustible, renewable energies were at that time envisioned as able to provide significant amounts of energy to solve short-term problems. In the meantime, heightened concern over environmental problems associated with some conventional energy sources has increased interest in renewable energy technologies whose environmental impacts have been considered relatively benign.

In such a context, most of the governments of IEA Member countries have initiated comprehensive RD&D and incentive programmes for increasing the pace of development and commercialization of these technologies. Although some of these government programmes and activities may have been hastily planned or sometimes "premature" for the stage of the technology targeted by the programme, the net effect of all the programmes was, nevertheless, acceleration of the development of these technologies and a greater understanding of the advantages and limitations of renewable energy sources and their technologies.

As a result of this experience, however, expectations concerning the pace of development and the contribution of renewable energies to energy supplies are now more realistic. This is partly because costs of conventional energies now seem unlikely to rise as far or as fast as was anticipated in the 1970s, and in fact, have recently fallen, affecting all alternatives. It is also because of a better assessment of the time needed for development and market penetration. These factors, combined with budgetary constraints, have weakened some governments' support and industry interest in developing alternatives to oil. The impact of these different effects leaves many countries re-examining their entire energy programme priorities and emphases, including of course renewable energy policies, R&D programmes and promotion activities. With their better understanding of these renewable energy technologies, governments are much better able to identify those that hold the most promise for them. This study attempts to provide the information and framework for beginning such an evaluation.

FINDINGS

Almost no generalities can be made about renewable energy technologies without a loss of accuracy or usefulness of the statement. There is not only tremendous diversity in the types of energy sources that are classified as renewable but also in 1) the amount of each resource that different countries have indigenously; 2) the technologies that utilise these energy sources and deliver useful energy to end users; and 3) the stage of development which each sub-technology has achieved. Although these sources vary greatly, they have a number of common characteristics, such as:

— independence from finite resources;

— wide geographical dispersion;

— low power density and for many, periodicity of supply;

— high capital cost but low operating cost;

— availability in small sizes, quickly installed;

— relatively low environmental impact;

— intensive manpower requirements.

Resource Assessments.

While the energy potential of renewable energy sources is very large, the nature of these sources is such that the energy is initially low in intensity and varies significantly with location. The geographic intensity of each type of renewable energy resource is fairly well known in IEA countries; however, specific locations of optimum sites have in many cases been less well identified. Appropriate resource "prospecting" techniques are well developed;although such testing is often lengthy and expensive. The extent of present and prospective use of some technologies could be wider if less costly techniques were developed, most notably in the case of identifying geothermal steam and hot water resources.

Technical and Economic Status.

The ability of a renewable energy technology to compete without support in energy markets is a primary determinant for predicting its success. The report finds that the technologies studied ranged from full economic competitiveness, even given today's lower oil prices and without government subsidies, to over ten times more expensive than conventional sources. To facilitate the discussion of the combined technical and economic status of this wide range of diverse technologies, the study divides the technologies into four categories.

1) "Economic" (in some locations): technologies which are well developed and economically viable at least in some markets and locations; further market penetration will require technology improvements, mass production, and/or economies of scale, such as

> Solar water heaters, replacing electricity, or with seasonal storage and for swimming pools
> Solar industrial process heat with parabolic trough collectors and large flat-plate collectors
> Residential passive solar heating designs and daylighting
> Solar agricultural drying
> Small remote photovoltaic systems
> Small to medium wind systems
> Direct biomass combustion power and heat applications
> Anaerobic digestion (of some feedstocks)
> Conventional geothermal technologies (dry and flashed steam power generation, high temperature hot water and low temperature heat)
> Tidal systems*.

2) "Commercial-With-Incentives": technologies which are available in the market but are competitive with the conventional technologies only with preferential treatments such as subsidies; these technologies still need further technology improvements, mass production, and economies of scale, such as

> Solar water and space heaters, replacing natural gas or oil
> Electricity generation systems with parabolic trough collectors
> Non-residential passive solar heating and daylighting
> Biomass liquid fuels (ethanol) from sugar and starch feedstocks
> Binary cycle geothermal systems.

3) "Under-Development": technologies which need more development to improve efficiency, reliability or cost so as to become commercial. This would include materials and systems development, pilot plants or field experiments to clarify technical problems and demonstration plants to illustrate performance capabilities and to clarify problems for commercialization, such as

> Solar space cooling (active and passive)
> Solar thermal power systems
> Photovoltaic power systems
> Large-sized wind systems
> Biomass gasification and energy crops
> Hot Dry Rock geothermal
> Wave energy systems.

---

* Only one tidal system, which was built in 1968, is presently operating commercially (in a non-IEA Member country). All other tidal projects are either proposed or in a demonstration stage, and would be considered to be "Under-Development" until constructed at commercial scale and proven "Economic" at a specific site under today's conditions.

4) "Future-Technology": technologies which have not yet been technically proven, even though they are scientifically feasible. Basic R&D on components would fit into this stage as would bench scale model development at laboratory levels to demonstrate the technical viability of the technology, such as

> Photochemical and thermochemical conversion of biomass
> Fast pyrolysis or direct liquefaction of biomass
> Biochemical biomass conversion processes
> Ocean thermal energy conversion systems
> Geothermal total flow prime movers
> Geopressured geothermal
> Geothermal magma.

Where appropriate, placement in the first two categories was determined by development of levelised present-value unit costs for both the renewable energy sources under consideration and conventional energy supplies for comparison. This approach to economic analysis of renewable energy technologies is necessary when comparing them to conventional sources because the former generally have high initial costs but low operating costs whereas many conventional sources are the opposite. For those technologies grouped under "Economic" or "Commercial-With-Incentives", this categorization must be used with care since the economic analysis included assumptions on resource intensity, conventional fuel costs, discount rates, technology efficiencies and energy output. All these assumptions can vary widely depending on many factors. Because of the variations in cost components in each country and for each renewable energy project, governments would benefit from performing their own analyses of renewable energy technology economics.

Renewable energy technologies in the "Economic" or "Commercial-With-Incentives" stages are generally more mature developmentally than those renewable energy technologies which have not reached these stages. These technologies have found market entry points where the resource is good and the applications are well suited to the specific technology. While some continuing R&D efforts are found to promise broader applicability of these technologies, the most beneficial effects for those in the first two categories would be 1) increasing production levels thus lowering the unit cost; 2) increasing the size of projects; and 3) ensuring where necessary that there are no institutional or other arrangements which produce a barrier to their commercial use on an economic basis.

For technologies in the "Under-Development" and "Future-Technology" categories the stage of advancement of the R&D for the technology, rather than present economic competitiveness with conventional fuels, is the determining factor used in this study for placement in the categories. Nevertheless, economic analysis is performed so that resource planners and policy makers will understand the order of magnitude of change in the economics that must be accomplished by technical improvements or other means before a technology can graduate from the pre-commercial stage. For those technologies in the last two categories, continued R&D successes are identified as the major need to move the technology into demonstration and commercial stages for those technologies in the "Under-Development" category and to determine the feasibility of the technologies still in the "Future-Technology" category. On-going and future R&D activities for each technology in all categories are identified in the report.

The geographic spread and site-specific characteristics of some renewable energy resources, the periodicity of others, together with the relative difficulties in transportation or storage, are found to present a limitation to expanded utilisation of some economic renewable energy technologies. These considerations present institutional challenges for full use of high resource sites and R&D challenges to develop better storage systems or the technologies themselves to the point where they can be economic in poorer sites.

Institutional Factors

IEA countries rely on market forces, and are, when necessary, supported by government actions, to produce a balance between energy sources. It is important that institutional arrangements which affect the energy market should as far as possible treat renewable and other sources of energy equally. This is most important when renewable energy technology projects are at the brink of crossing the threshold of economic viability and would be most assisted by a smooth path toward introduction to the market. There are at present a number of inadvertent and widespread biases which may prevent these emerging technologies from achieving significant market penetration in a timely fashion. These include the barriers common to any new technology which is being introduced into the market, such as a lack of 1) standardization of systems; 2) uniform approval practices; 3) public understanding of the probable impacts or benefits; and 4) infrastructure, and, as a consequence, 5) conventional financing.

There are also, on the one hand, more specific impediments such as energy prices which are inevitably affected by short-term influences and may not reflect the long-term outlook. Another impediment is a tendency to require higher rates of return on investment in renewable than in conventional sources of energy. Electric utilities play a key role in the development of some renewable energy sources; in the past they may sometimes have used their market position to discourage competition from renewable energy technologies. On the other hand, there have been cases where government policies, such as the Public Utility Regulatory Policy Act of 1978 in the United States, had the effect of opening electricity markets to generating from renewable energy sources and other small power producers.

Government Support

Where national policy objectives are met by development of renewable energy technologies, government support to accelerate the development and/or commercialization of these technologies can be appropriate and can take the form of a number of activities, ranging from R&D programmes to demonstrations which contribute to the development of the technology base for the renewable energy technologies.

Government activity in funding R&D for renewable energy technologies has been very significant, over US $7 billion cumulatively by the IEA countries through the end of 1985. Many technical problems have been solved, a number of uncertainties reduced, and solid technical bases established. This support and its duration have had the combined effect that renewable energy technologies have been continually advancing technically, increasing their contribution to meeting some countries' energy requirements and contributing to the goals of diversifying their energy supplies.

There have been recent reductions in some R&D budgets for renewable energy technologies. Some technologies have become economic and no longer need as intense government R&D efforts; others have been determined to be of lower priority or less appropriate to national energy needs. General reductions in all energy R&D reflecting financial pressures have affected renewable energy technology funding as well. Some countries have integrated renewable energy technology policy with industrial policy and sometimes economic development activities because of renewable energy technologies' spin-off benefits in these other areas.

Government incentives, data bases, information programmes, regulations and leadership activities also have allowed an array of renewable energy technology industries to emerge sooner than they would have without this government involvement. In some cases, such as small wind systems, government support has moved the technology from the R&D stage into the market-place. This process has further helped create the commercial capabilities of some renewable energy technology industries; although it has sometimes been inefficient. It has therefore paved the way for governments to understand how to assist emerging renewable energy technology industries in the future with greater efficiency.

While much has been learned from the various government activities, the key to maximum effectiveness is timing. Government support is most effective when carefully tailored to the stage of development of the particular technology. For example, during the R&D phase support might go to research institutions and industry; whereas during technical and commercial demonstration support might go to industry because it can provide the required technology push and multiplication. In the last phase, to promote market pull and commercial penetration, government support might go to purchasers.

Finally, it is noted that the development of renewable energy technologies is no different than other major energy sources where commercialization can take decades. On the one hand, inappropriate government policy changes or decreased commitment in response to short-term factors can produce major set-backs to important long-term programmes. On the other hand, continued support where unwarranted by technical advances and economic outlook can also be harmful. Consistency of government policy supported by programme reviews of progress is most helpful.

Opportunities for collaboration

International collaboration on R&D for promoting renewable energy technologies has taken several forms. Most IEA countries participate in one or more bi- or multi-lateral agreements to share information and results from R&D on similar technological developments; many participate in joint renewable energy technology projects. These co-operative activities have occurred through several countries' own initiatives and through initiatives of international organisations such as the IEA itself or the European Community which has substantial RD&D programmes. These activities have not only enhanced the information available on renewable energy technology R&D and increased the pace of development, but they have also decreased the cost of R&D for individual countries.

Conclusions and Recommendations

The contribution of renewable energy technologies to energy supplies in IEA countries is, at present, relatively small because they are competitive with conventional sources only in specific markets. The technical development of renewable energy technologies has, however, made remarkable progress during the past twelve years, and, for most of them, the outlook for continued improvement is good. They are likely to make an increasing contribution as an addition to existing energy sources. In the meantime, however, as with any new energy source, even in a climate of substantial support for technology development, at least 30 years may be needed for them to achieve a significant market penetration, which will, nevertheless, still be relatively small in terms of share of total energy supply in IEA countries. In the meantime RD&D investment may provide profitable returns because the world energy market is so large that even a relatively small penetration may provide good business opportunities for a technology.

Although predictions are notoriously difficult, governments can establish priorities for their involvement in renewable energy technology development, based on technical merit, potential supply contribution, economic competitiveness and other important factors such as indigenous availability, degree of diversification and potential environmental advantages. Renewable energy sources, like other sources of energy that a government wishes to promote, will undoubtedly require long-term government RD&D support;

and, while criteria for setting priorities will certainly differ among governments, it is important that they are developed and that periodic re-evaluations of the economic and technical status of the chosen priorities are carried out. These re-evaluations can raise the quality of those early analyses which were necessarily limited by the reliability of the data available, and at the same time help identify technologies which have moved from the earlier stages to "Economic", thereby eliminating the need for continued government support. Their quality would be further improved by a better understanding of the economics of these technologies, taking into account such benefits as security of supply, environmental advantages and locally-produced energy. Governments may also need to influence the institutional and commercial environments in order to facilitate market penetration of successfully demonstrated technologies.

International organisations working on renewable energy technology collaboration are contributing to the base of knowledge available on renewable energy technologies. It is especially important for these organisations, including the IEA itself, to co-operate wherever appropriate, to eliminate needless duplication and ensure a better use of scarce resources. The private sector could be encouraged to become more involved in the development of renewable energy technologies, especially at the international level, with due consideration given to protection of their commercial interests and proprietary information.

The specific recommendations for consideration by governments are:

—   maintain long-term RD&D efforts, supported by evaluation of past government programmes and activities and of the economic and technical progress of the technologies targeted, since continuity is indispensable for introduction of new renewable energy technologies into the market, and seek broader participation of the private sector in government-sponsored collaborative activities;

—   review priorities for further development of renewable energy technologies, focusing on those which seem most promising, considering each country's local resources and national energy policy constraints and outlook;

—   encourage the development of an economically-viable renewable-energy-technology industry and the development of innovative financing arrangements for economic investment in renewable energies;

—   seek, as a matter of general energy policy, in accordance with agreed IEA objectives, to avoid and remove distortions in the energy market, particularly energy prices and criteria for the assessment of investment;

—   encourage, either directly or through their influence with the appropriate regulatory authorities, electric utilities to base their policies for buying and selling electricity to renewable energy producers on avoided costs;

— carefully plan and design policy actions supporting the development of renewable energy technologies, such as the adoption of standards and codes or the modification of utility regulations, the development of data bases and reliable and appropriate information for potential users and installers of renewable energy devices, financial incentives and the removal of institutional barriers which inadvertently block the commercialization of renewable energy technologies; and

— concentrate international collaboration on:

- development of specific components that are preventing advancement of technology;

- workshops on the problems of commercialization and on economic analysis of renewable energy technologies;

- stimulation of renewable energy resource "prospecting" by private industry, and work on developing a common methodology to estimate future contributions from renewable energy sources; and

- technology transfer among countries and from governments to industry.

CHAPTER I

INTRODUCTION

The use of renewable energy sources is as old as man himself. Even today much of the world's population still relies on firewood for heating and cooking. Some renewable energy technologies, such as large hydro-electric generation and wood-fired boilers for steam and electricity generation, have been proven technologically and commercially for a long time. But commercial and industrial interest in developing technologies for utilizing renewable energy resources such as solar, geothermal and ocean thermal and ocean kinetic energy is more recent. It was only after the substantive oil price increases a decade ago that the public and private sectors in IEA countries made a committed effort to develop these energy sources.

This study takes a look at renewable energy sources in IEA Member countries and their related technologies in order to assess:

— the accomplishments and lessons of more than a decade of developments;

— the most promising areas of further RD&D and/or other government action to increase the development and utilization of renewable energies; and

— the outlook for contributions to national energy supplies on an economic basis.

The ten technology areas discussed in this report are:

— active solar energy;

— passive solar energy;

— solar thermal energy;

— photovoltaic energy;

— wind energy;

— biomass energy;

—   geothermal energy;

—   tidal energy;

—   wave energy; and

—   ocean thermal energy.

Hydropower is not discussed because, although there is some scope for innovation and promotion, particularly at the smaller scale, it is felt that this resource and technology are sufficiently well-developed and commercially available to obviate the need for further investigation. Hydrogen production has been omitted because hydrogen is not a primary energy source.

Chapter II summarises the present status and prospects of each of these renewable energy technologies and provides a technical and economic framework for classifying the diverse number of renewable energy technologies. The chapter is supported by more detailed analyses in Annexes I and II. Annex I provides examples of the technical status and outlook of renewable energy technologies, while Annex II provides an assessment of the economic status of each renewable energy technology as compared to that of conventional energy sources. Chapter III discusses institutional and other factors which affect the development of renewable energy technologies. Chapter IV describes government activities in support of the development of renewable energy sources and seeks to assess the strengths and weaknesses of these activities. Chapter V examines the progress of international collaboration and the scope for its development. Chapter VI seeks to draw policy conclusions from the earlier analyses and to suggest ways in which progress can be made at both national and international levels.

BACKGROUND

In 1974, when many IEA Member countries were initiating their renewable energy technology planning and programmes, most of the technologies discussed here were either in the conceptual stage or existed in "first generation" forms. Only large hydroelectric generation and wood-fired boilers for steam and electricity generation had been proven technologically and used commercially for a long time. Solar water heaters and biogas digesters were available but in rather limited use. Photovoltaic systems, on a very specialised and small scale, had already been used in space. Small wind machines to pump water and provide remote electrical output had been available for decades, and several dry steam geothermal plants were in commercial operation. Use of woody biomass for space heating and production of process heat and electricity, especially in wood processing industries, was well established. Small wave power generators had been used for navigational buoys since the middle of the 1960s.

Renewable energy technologies, like the energy sources they are associated with, vary enormously. Therefore, renewable energy sources and technologies cannot be lumped together under one homogeneous and coherent heading or discussion. But there are a number of characteristics in common, stemming from the similar nature of renewable energy sources, such as the following:

—   renewable energy sources are those which, in terms of the human life span, are perpetual and therefore virtually inexhaustible; these include solar radiation, atmospheric winds, ocean thermal and ocean kinetic energy, geothermal energy, thermal and chemical energy from biomass, and mechanical energy from falling water;

—   renewable energy sources are widely dispersed and, in general, for either physical or economic reasons, they are developed and used at sites where the resource is available;

—   renewable energy sources generally have low power and energy density and, as a result, their technologies often have large space requirements;

—   renewable energy technologies often have high initial costs, while operating costs are low and relatively insensitive to fluctuations in conventional energy prices;

—   renewable energy technologies are often modular and can be installed relatively quickly in response to varying demand; and

—   renewable energy technologies are generally environmentally benign; most of them are unlikely to cause significant public acceptance problems; however, with widespread use of certain technologies (such as wood combustion), the possible environmental impact could affect public acceptance.

Major efforts have been made in IEA countries to develop renewable energy technologies since 1974. From 1977 to 1985 Member governments of the IEA have devoted over US $7 billion (in 1985 prices) to research, development and demonstration of renewable energy sources. Substantial additional support in the form of grants, favourable loans and tax incentives has also been given to encourage the use of certain developing renewable energy sources. Over the same period, industry has invested a similar amount of money in renewable energy projects.

Much successful research, development and demonstration work has been accomplished to develop renewable energy technologies, from the design of photovoltaic cells to the development of geothermal drilling techniques. New materials have been developed with the thermal, optical and physical properties required for lower cost and higher performance. New components have been tested and substituted for less effective ones. Extensive systems analysis has been performed with thousands of systems studied under both controlled laboratory conditions and typical field operation conditions. Throughout IEA Member countries, teams of scientists and engineers in universities, industry and government laboratories have created a solid technology base and expertise in the extremely broad array of disciplines that make up renewable energy technologies.

Renewable energy sources are now used in all Member countries of the IEA, and, in some of them, these sources contribute significantly to the national energy supply. In all IEA countries their contribution to the energy supplies is growing as R&D efforts are reaching fruition and market penetration increases. Renewable energy (excluding hydroelectric generation) provides between 1% and 5% of total primary energy requirements in Australia, Austria, Canada, Denmark, Sweden and Switzerland. In Ireland, peat resources provide about 8% of the country's primary energy requirements and in Portugal biomass resources cover about 7% of those requirements. In absolute terms in the United States, the energy contribution from renewable energy sources amounts to about 64 million metric tons of oil equivalent per year, which is roughly equivalent[1] to the total energy requirements of the Netherlands.

On the other hand, compared with the early 1970s, expectations concerning the pace of development are now more realistic. This is partly because fuel costs, far from rising as fast as expected, have in fact recently fallen drastically, affecting all alternatives. It is also because of a better assessment of the time needed for development and market penetration. These factors, combined with budgetary constraints, have weakened some governments' support and industry interest in developing renewable energy technologies. The combined impact of these different events leaves many countries re-examining their entire energy programme priorities and emphases, including of course renewable energy promotion activities, policies and R&D programmes.

CHAPTER II

RENEWABLE ENERGY TECHNOLOGIES DEVELOPMENT AND OUTLOOK

For any energy source and its related technologies there are several factors which establish its outlook for a country:

— the available resource base;

— the technical status of the technology used to turn the resource into usable forms;

— the type, and quality of energy supplied; and

— the technology's commercial status (economics, institutional barriers and market factors).

This chapter broadly summarises the combined major findings of Annexes I and II which might be useful to policymakers in determining future renewable energy technology programme directions mainly as they relate to the four factors listed above.

RENEWABLE ENERGY RESOURCE BASE

Accurate renewable energy resource assessment is important on two levels: 1) for governments to assess the potential contribution to national primary energy requirements and 2) for specific renewable energy projects to be economically successful. Most IEA countries have done national renewable energy resource assessments on the first level in the last ten years. These assessments have varied tremendously in their sophistication, methodologies and time periods. For example, some countries have completed detailed wind atlases using hourly anemometer measurements in key locations over several years. Others have assembled wind information from existing weather stations, which are not necessarily located in areas of highest wind potential. For this reason, neither summations of resource assessments nor cross-country comparisons are provided here.

Where particularly good examples of resource assessments exist, these are noted in Annex I in the discussions of renewable energy sources. This includes the aspects of the resource which are the most important for the technology applications that have been or are being developed. In general, however, all IEA countries have been found to have solar energy, wind energy and biomass resources to varying degrees. Accessible geothermal energy and ocean-based resources are highly geographically-specific and not available in all IEA countries. For example, while all IEA countries except three (Austria, Switzerland and Luxembourg) have coastlines, conditions are not always favourable enough to warrant wave machines. Nevertheless, the combined potential of all renewable energy resources is thought to be enormous and virtually untapped even after twelve years of rapid technology development. Summarizing some examples of assessments of these potential resources as provided in Annex I gives a rough indication of the magnitude:

— active solar, between 1% and 2%, and biomass, over 7%, of total energy requirements in all IEA countries;

— geothermal, hundreds of thousands of MWh equivalents, and

— wind, 2 million GWh annually in European countries.

Site-specific resource assessment is one of the more critical factors in determining the successful application of any specific renewable energy project. This second level of assessment becomes most important when technologies have become commercialized, and specific sites are being identified. The technologies which are most sensitive to resource assessment for their continued expanded application and the problems associated with this assessment are discussed in Annex I and are identified later in this chapter in the paragraphs discussing each technology individually.

COMBINED TECHNICAL AND ECONOMIC STATUS

As mentioned in the introduction, two factors important for introduction of a technology into the market are its technical status and its commercial status. While the Annexes cover these two aspects separately, this discussion combines them in an attempt to provide an integrated summary of each technology's status. In order to ensure that the terms used are consistent and clear to the reader, the following references to development stages will be used throughout the report:

1) "Economic": technologies which are well developed and economically viable at least in some markets and locations; further market penetration will require technology refinements, mass production, and/or economies of scale;

2) "Commercial-With-Incentives": technologies which are available in some markets but are competitive with the conventional technologies only with preferential treatments such as subsidies; these technologies still need further technology refinements, mass production, and economies of scale;

3) "Under-Development": technologies which need more research and development to improve efficiency, reliability or cost so as to become commercial; this would include materials and systems development, pilot plants or field experiments to resolve operational problems and environmental impacts and demonstration plants to illustrate performance capabilities and to establish cost and performance capabilities of specific applications.

4) "Future-Technology": technologies which have not yet been technically proven, even though they are scientifically feasible; applied R&D on components would fit into this stage, as would bench scale model development at laboratory levels to establish the technical viability of the technology.

Generally, for a technology to be classified under one of the above stages, most of its recent applications should have reached that stage, although a technology would still be considered "Under-Development" in cases where only the first commercial scale project has been built to demonstrate the commercial viability. Where there is more than one technology in a technology area at different stages, the technologies are subdivided. The distinction is made between "Economic" and "Commercial-With-Incentives" because (as will be shown in Chapter IV) over the last ten years or so, various subsidies and incentives have been available for some

TABLE 1
CURRENT STATUS OF RENEWABLE ENERGY TECHNOLOGIES
(in some locations)

ECONOMIC

Solar water heaters, replacing electricity, or with seasonal storage and for swimming pools
Solar industrial process heat with parabolic trough collectors or large flat-plate collectors
Residential passive solar heating designs and daylighting
Solar agricultural drying
Small remote photovoltaic systems
Small to medium wind systems
Direct biomass combustion
Anaerobic digestion (of some feedstocks)
Conventional geothermal technologies (dry and flashed steam power generation, high temperature hot water and low temperature heat)
Tidal systems *

COMMERCIAL-WITH-INCENTIVES

Solar water and space heaters, replacing natural gas or oil
Electricity generation with parabolic trough collectors
Non-residential passive solar heating and daylighting
Biomass liquid fuels (ethanol) from sugar and starch feedstocks
Binary cycle hydro-geothermal systems

---

* Only one tidal system, which was built in 1968, is presently operating commercially (in a non-IEA Member country). All other tidal projects are either proposed or in a demonstration stage, and would be considered to be "Under-Development" until built and proven "Economic" at a specific site under today's conditions.

TABLE 1 (Continued)
CURRENT STATUS OF RENEWABLE ENERGY TECHNOLOGIES
(in some locations)

UNDER-DEVELOPMENT

Solar space cooling (active and passive)
Solar thermal power systems (other than parabolic trough collectors)
Photovoltaic power systems
Large-sized wind systems
Biomass gasification
Hot dry rock geothermal
Geothermal total flow prime movers
Wave energy systems.

FUTURE-TECHNOLOGIES

Photochemical and thermochemical conversion
Fast pyrolysis or direct liquefaction of biomass
Biochemical biomass conversion processes
Ocean thermal energy conversion systems
Geopressured geothermal
Geothermal magma.

---

renewable energy technologies in several countries. These subsidies have been quite instrumental in bringing certain technologies to the market, while others still have not made sufficient technical improvements to be fully economic without the subsidies. For each technology studied, Table 1 illustrates the current status by these combined technical/economic classifications.

Table 2 provides a breakdown of the renewable energy technologies by typical size, type of energy supplied and applications. These two tables summarise the findings of Annexes I and II relating to the technologies' present status. For detailed information on each technology and its outlook, the reader must refer to Annex I. Chapter V has a summary table of the findings of Annex I on R&D needs by technology. The factors that determine in which stage a technology is placed are described below with emphasis on characterising the present level of development and describing the main technical or economic factors that must be dealt with to substantially change a technology's status. The economics are presented as rough comparisons, in orders of magnitude, with conventional fuel costs because economic analyses for renewable energy technologies are so site-specific that it would be misleading to state costs per kilowatt hour here without a full description of the assumptions used in the analysis. These assumptions are, however, fully described in Annex II.

While the discussion which follows draws on Annexes I and II, here the subsections are organised according to groups of technologies in the four levels of development: "Economic", "Commercial-With-Incentives", "Under-Development" and "Future-Technology". In this way, the characteristics, such as the types of barriers still remaining, are roughly equivalent, and the discussions of them will be more coherent.

TABLE 2
RENEWABLE ENERGY TECHNOLOGIES, SIZE RANGES,
APPLICATIONS AND TYPE OF ENERGY SUPPLIED

TECHNOLOGY AREA AND TECHNOLOGIES (AND TYPICAL SIZE RANGE)	APPLICATIONS AND TYPE OF ENERGY SUPPLIED
Active solar (1-500 kW$_{th}$) Flat-plate collectors Evacuated-tube collectors	Water heating Pool heating Space heating Space cooling Industrial process heat
Passive Solar (1-500 kW$_{th}$) Integral design Specialised components	Space heating Space cooling Daylighting
Solar Thermal (30-100 MW$_e$) Parabolic troughs Parabolic dishes Central receivers Solar ponds	Industrial process heat Electricity Irrigation pumping Desalination Liquid fuels and chemicals production
Photovoltaic Energy (up to 6.5 MW$_e$) Single-crystal silicon cells Polycrystalline silicon cells Polycrystalline silicon ribbon Thin Films	Electricity for grids and remote system Consumer electronic products Concentrator Cells Space applications
Wind (up to 500 MW$_e$) Small to medium scale Large scale	Electricity Irrigation Pumping Heat
Biomass (up to 265 MW$_e$) Anaerobic Digestion Fermentation/Distillation Chemical Reduction and Distillation Pyrolysis, Liquefaction, Gasification Hydrogenation Direct Combustion	Heat Electricity Liquid and gaseous fuels

Geothermal
Hydrogeothermal (up to 50 MW$_e$)
- Dry and flash steam
  Binary cycle
- Total flow prime movers
  Hot Dry Rock
Geopressured
Magma
	Electricity
	Direct Heat
	Methane production
	(geopressured only)

Ocean Wave (3-500 kW$_e$) - Surface followers - Pressure activated - Focusing devices Tidal (240 MW$_e$) OTEC (1-40 MW$_e$) - Closed-cycle - Open-cycle	Electricity Mechanical energy Electricity Electricity Aquaculture Fresh water production

Economic Renewable Energy Technologies

Active solar water, swimming pool, space heating and industrial process heating systems are presently economic primarily in areas with good solar energy resources and high fuel costs. Systems serving larger energy demands tend to be more economic. Millions of simple solar water heating systems have been sold in IEA countries, mainly in the sunniest regions. Fewer space heating systems have been installed. Residential installations account for the majority, but many successful commercial installations exist as well. The widespread requirement for hot water and space heating in both sectors and the general availability of solar energy throughout IEA countries indicates a significant potential for further penetration of the technology. R&D is concentrating on improving efficiency reliability and durability of components, reducing the materials-intensiveness of solar collectors with new materials and improving energy storage capabilities, especially for space heating. Techniques being developed to simulate the performances of complex building systems will be useful for large solar heating systems. Long-term inter-seasonal storage added to solar heating systems is being developed to the point where it may be able to increase significantly the amount of insolation utilised and therefore greatly expand the region of economic applicability of these systems. Large-scale, long-term energy storage is already at the demonstration stage in Sweden. If these efforts are successful, the number of economic applications would increase substantially.

Passive solar designs, combined with conservation techniques, can economically provide a large share of the energy needed for residential space heating, and many cost no more than conventional designs. Passive solar can be adapted to a wide range of climatic conditions; thus the potential for the application is great in most IEA countries. The extent of present diffusion of the use of these techniques cannot be known exactly but is felt to be a very small fraction of its economic potential. Diffusion is limited by unadaptable existing building stocks and by lack of solar access. R&D efforts are concentrating on improved materials and specialised components which could increase the number of economic applications (e.g. where solar access does not allow a simple passive solar design). However, dissemination of information on the technology to the public and building industry is most important for increasing the penetration rate of this technology.

Daylighting, for reducing lighting and cooling energy use in non-residential structures, again, in most cases adds little or nothing to construction costs and yet the techniques have only relatively recently been introduced and therefore are only being used to a limited extent. R&D efforts in the area of core daylighting and better system controls should increase the number of economic applications, while information dissemination efforts should increase the rate of adoption of the techniques by the building industry. Significantly extending the applicability of passive space heating and daylighting in non-residential structures requires better understanding and designs of complex systems and better building materials.

Wind technologies are fairly advanced for *small- and medium-sized installations* (up to 500 kW). Small- and medium-size wind power systems offer a promising alternative in high resource areas, such as some coastal areas or areas with unusually high wind conditions. Over 10 000 wind machines have been installed in the last decade, the bulk of these in "windfarms"; although only recently have such installations begun to be economic without subsidies. Development of small-to-medium size turbine technology is relatively mature; however, this size of wind system could benefit from mass production. Further R&D should concentrate on improving wind capture potential, reliability, life expectancy and availability to substantially reduce costs and enhance performance, thereby increasing the number of sites where wind systems would be economic. More accurate site-specific assessments of wind energy potential would spur market interest in countries and regions where there is potential but little market penetration to date.

Biomass technologies for providing *heat and combined heat and power plants* based on wood, peat and municipal wastes are in widespread use with many applications. Hundreds of small-scale systems for electricity generation, industrial process energy and district heating with wood, peat, refuse-derived fuel, industrial waste and straw are presently in operation in numerous IEA countries. These plants are especially attractive for agricultural or forestry industries since biomass transportation costs are minimised. Wood stoves for residential heating are widely used, in fact, so much so that in some areas they are causing concern for environmental degradation of air quality caused by emissions. R&D is being conducted to develop energy crops which increase biomass supplies, to densify biomass fuels, and to develop lower cost collection and handling equipment. The development of the energy crop concepts will involve government policies regarding forests, agriculture and land use. Success in these areas could significantly extend the number of economic applications. R&D on air emission control technologies and solid waste disposal techniques could also overcome barriers to some projects in areas where there is sufficient high grade fuel but where air pollution and waste disposal limitations exist.

Anaerobic digestion to produce methane is widely used in many IEA countries. This use is increasing because of process improvements and advances in the feedstocks which can be utilised and in feedstock pretreatment. Dissemination of information on how and where this technology is applicable and economic could further expand the use of this technology. This technology could be improved and the types of feedstocks expanded by research in basic biochemical and microbiological reactions, material durability, process engineering and pretreatment of feedstocks.

In some of the best geothermal fields, costs for *electricity generated by geothermal hydrothermal* technologies (using high temperature steam or flashed hot water) are half the cost of oil-fired generation and three-quarters of the cost of nuclear or coal generation. At the end of 1986 approximately 5 500 MW_e will be operating worldwide. New plants are being constructed, and some existing installations are being expanded. Nevertheless, to realise the full potential for using these technologies to convert geothermal energy, progress must be made on techniques to assess reservoir potential in order to reduce the financial risk associated with geothermal resource development.

In addition, *low-grade geothermal heat* contained in warm and hot water aquifers is being exploited economically for *space heating and district heating.* Most uses of the technology where the resource is adequate are less costly than or equal to oil heating costs. At the end of 1984 a capacity of over 7 000 MW$_{th}$ was installed worldwide. The primary limitation to increased use of geothermal direct heat is geographical. Matching industrial process heat requirements to geothermal sources requires locating industries at resource sites which are often remote locations.

Tidal power systems are economic when compared to total cost of conventional oil generation but still almost twice as expensive as coal-fired generation. However, there is presently only one tidal energy power station in operation. Because the technologies for construction of tidal systems are well developed, economy of scale is one of the most important factors for the utilisation of tidal energy. Tidal power projects are therefore likely to be quite large in comparison with other renewable energy projects. Proposals for tidal power projects estimate a required capital investment in excess of a few billion dollars. Very few enterprises in the private sector could carry out such a project even when it promises to be very profitable. Therefore, government assistance or participation to mitigate the risks of the large initial capital investment seem to be indispensable for the development of tidal energy.

"Commercial-With-Incentives" Renewable Energy Technologies

Two *solar thermal* technologies, *parabolic troughs* and *solar ponds* for the production of industrial process heat are much further advanced developmentally than the other solar thermal technologies. However, only a few systems had been installed before tax advantages in the United States were eliminated and oil prices dropped. Improved manufacturing economies would make the most impact on increased penetration.

Conversion of biomass to ethanol using sugar and starch feedstocks has reached large-scale proportions in the United States and Brazil where the technology is receiving government support and the equivalent of hundreds of millions of barrels of gasoline have been produced. Nevertheless, the cost remains 1.5 to 3 times more expensive than conventional fuels. Continued R&D in celloluse conversion and utilisation could provide the cost reductions necessary to make ethanol production fully economic, possibly as soon as 1990.

Solar water and space heating replacing natural gas or oil heating systems has experienced some market penetration in Japan without incentives, but elsewhere with the assistance of government incentives, which offset the added cost (presently up to 1.5 to 2 times more expensive than the alternative). This industry has declined significantly in countries where these incentives have been removed.

Binary cycle plants to utilize liquid-dominated geothermal resources could expand the potential for geothermal electricity production by allowing use of lower temperature liquid-dominated geothermal sources and may be useful in hot dry rock geothermal resource utilisation as well. Several commercial plants are operating in the United States where a 45 MW$_e$ proof-of-concept binary plant has also been built and is being monitored. R&D efforts are concentrating on increasing performance and decreasing cooling water requirements.

"Under-Development" Renewable Energy Technologies

Active solar space cooling technology still requires substantial R&D despite many advances in the technology to date. R&D is concentrating on increasing the efficiency of energy conversion equipment, such as desiccant-dehumidification cooling, absorption cooling, and hybrid systems. With success in making these improvements, the efficiency of solar cooling systems is expected to increase by a factor of two which could make them economic in good resource areas by the year 2000.

Passive solar space cooling techniques lag behind passive solar heating techniques in terms of research and use. Their usefulness also appears more limited since they are more dependent on local climatic conditions. Use of this technology in all types of structures still needs R&D emphasis.

A number of *solar thermal technologies* (parabolic dishes and troughs, central receivers and solar ponds) are based on different system concepts for receiving and converting the solar energy to useful energy. A decade of development has brought the specific costs of present systems for electricity generation to a factor of two to six times higher than conventional alternatives except for troughs which are only two to four times higher. Results from pilot and/or demonstration plants indicate that a variety of R&D activities are needed to significantly improve the performance and reliability of these solar thermal systems. Two important R&D areas are development of improved methods of storage and transport of chemicals. In addition, mass production of components and lower operation and maintenance costs will be necessary. This further technology development is expected to lower costs by 50% or more which would make these solar thermal technologies attractive to the high value electricity market in the late 1990s. Finally, better understanding of system lifetimes and of optimising solar applications to satisfy total loads would be necessary to attract capital for projects once the technology has been shown to be economic.

Photovoltaic electricity generation development has succeeded in bringing generating costs to around ten times the present costs of conventional generation. Reaching the economic stage is projected to occur by mid-1990s. Only remote applications and consumer products have seen any market penetration, and in 1984 world-wide photovoltaic shipments totalled about 25 MW. Photovoltaic technologies development currently is taking a variety of approaches in order to produce more efficient cells at less cost. These approaches include development of different photovoltaic materials and production techniques (single crystal and polycrystalline silicon, amorphous silicon, thin films, concentrator cells), use of concentrators, and reducing costs of the balance-of-system components. Lifetimes of systems are still a major unknown and would have to be firmly established before large systems could be financed conventionally.

Demonstrations of *large-scale wind systems* have proved to be three to six times more expensive than conventional generation. Many technical problems must be overcome by R&D in the areas of blade design construction and testing, aero-dynamics and atmospheric physics and materials. Other areas of R&D which could enhance the potential of both large and small systems are: development of load following diesel-hybrid systems, offshore wind energy stations, and techniques for improving interconnection with electric grids to reduce surge problems. Predictions of when this technology may become economic are risky because, on the one hand, the pace of R&D efforts has slowed substantially in some countries, while, on the other hand, some countries see relatively large potentials for contribution from this technology (e.g. Germany: 1.3% to 1.8% of electricity requirements by the year 2000).

In the area of *biomass gasification,* several technologies are being developed, and low BTU gasification techniques have reached demonstration stages. In general the costs presently exceed the cost of conventional sources by a factor of 1.5 to 2.5; although this technology is marginally competitive in some agricultural applications. Gasification R&D needs to address process problems such as those that interrupt smooth operation of such systems. These problems can be relatively easily solved; so movement of this technology into the market place is likely to be accomplished soon.

Testing of a geothermal total flow prime mover has been completed in several countries. Recent Hot Dry Rock technology developments have shown that reservoirs can be created and that this technology can produce hot water. Progress is still needed in demonstrating the heat capacity of the reservoirs.

A variety of designs for *wave energy conversion devices* are being developed and tested. Some experience with costs of these types of systems indicate that they are around twice as expensive as diesel generation. Full-scale testing is occurring in two recent, privately-constructed projects in Norway with projections of costs being close to competitive with conventional generation. R&D is concentrating on ways to improve energy conversion efficiency and system design, integration and components.

"Future-Technology" Renewable Energy Technologies

Biomass technologies such as *fast pyrolysis, direct liquefaction and biochemical conversion* are being pursued because of the large biomass resources available and because they promise to produce liquid and gaseous fuels and energy-intensive petrochemical substitutes. If current research efforts are successful, some of these products could be available commercially around the year 2000. Technologies to tap *geopressured geothermal* and *magma* sources of geothermal energy are being developed in order to greatly expand the amount of resource which is considered economically useful. The very large potential offered by these energy sources is offset, at least initially, by the high costs of developing the technologies. Magma R&D is at the level of development of equipment and materials capable of withstanding the very high temperatures that will be encountered. Geopressured geothermal resources development requires combining systems to capture dissolved methane with others to extract heat. R&D in this area is also concentrating on reservoir engineering and prediction techniques. *Ocean thermal energy conversion technologies* are being developed because the potential resource is very large. R&D efforts in materials and components development are expected to lead to a proof-of-concept experiment of significant size.

Type of Energy Supplied by Renewable Energy Source Applications

Consideration of the type, quality and location of energy supplied by renewable energy technologies helps to understand the importance of specific renewable energy technologies and their applications to energy markets and to the energy needs of a country. Most of these technologies readily produce, and, therefore, have capabilities to contribute to, electricity generation supplies; however, electricity from solar, wind, tidal and wave energy sources is either periodic, intermittent or both. Periodic electricity is where the supply is interrupted at regular intervals (such as tidal power), and intermittent electricity is where the supply is interrupted unpredictably (such as wind energy). The supply patterns from intermittent or periodic sources do not necessarily match users' needs; therefore, renewable energy systems for these sources often include energy storage devices or back up systems based on conventional energy sources. Recently some success has been obtained with systems which couple diesel or gas turbine generators with intermittent sources (hybrid) in order to reduce the impact to electricity grids and stand-along systems of intermittent supplies. However, additional R&D would be advantageous to develop devices to achieve full systems integration of intermittent electricity sources with electricity grids and better storage systems.

Other renewable energy technologies (e.g. geothermal, biomass, ocean thermal) can be designed to provide baseload or peaking electricity according to user needs. Several renewable energy technologies (solar, wind, biomass, and geothermal) are best used to supply energy for agricultural and industrial process heat and water and space heating. Of course a number of these technologies initially produce mechanical energy which can be used as such; although electricity represents a premium application and is therefore produced instead. Only biomass holds near-term promises of producing liquid fuels, such as those used for transport; although solar thermal and geopressured geothermal are believed to have potential to produce liquid fuels in the long term.

Careful planning and systems analysis to fit the energy supplied into the prevailing energy system can optimise the contribution of renewable energy technologies. For example, tidal generation is well suited to pumped storage of water to produce peak hydro-electricity; solar thermal electricity generation to meeting peak cooling loads; and active solar heating to swimming pool heating and agricultural drying.

As mentioned in the previous discussion of the status of these technologies, some are quite geographically specific. High resource sites are often not where the energy produced is most needed. Those technologies producing electricity can be transported relatively long distances to users; however, the sometimes significant additional cost of new transmission capacity must be considered in decisions to develop a resource site. For those technologies producing energy in "non-transportable" forms such as heat, project sizes are restricted to the energy requirements of the users at the location of the resource; the economy-of-scale needed to make the project viable may sometimes not be realisable. Location of users near high resource sites may not be realistic until the value (and reality) of renewable energy as a lower-cost energy option is recognized by both the public and decision-makers.

Economic Analysis

The economic analysis is meant to serve two purposes. First, for renewable energy technologies that are "Economic" or "Commercial-With-Incentives", economic analysis can provide an indicator or "yardstick" of whether the technology can compete successfully with conventional energy sources and hence whether it can be expected to supplement them without further assistance or R&D effort. (How much the technology would supplement the conventional sources would be a function of resource availability and geography as well as many institutional factors.) Second, for technologies that are still in the research and development stage, economic analysis helps understand the order of magnitude of technical improvements (the cost-reduction outlook) that would have to be made before a technology would begin to be economic. (How quickly this would happen is a function of the type of technical improvement and how much effort is devoted to achieving it.)

In general only those technologies which are in the "Economic" or "Commercial-With-Incentives" stages can be the subject of detailed economic analyses. In other words, the data on other technologies which have not yet reached that stage, are still insufficient and may not fully reflect a technology's future potential. Great uncertainties exist in emerging renewable energy technologies concerning potential performance improvements and the possibilities of cost reductions. Therefore, descriptions of the economics of those technologies not yet in the "Economic" or "Commercial-With-Incentives" stages should be used with care. Also, economic calculations should be carried out on a periodical basis in order to reflect new technical achievements and practical experience gained. Finally, it should be realised that not all factors can be taken into account in economic analysis. These may be just as essential as economics for the future successful contribution of a renewable energy technology. These factors are discussed in the following chapter.

Policy-makers need to know the status of a technology without preferential treatment so as to gauge whether any government activities are warranted, especially policy-makers who are not completely familiar with all the conditions that surround technology development in another country. Therefore, for this study, the economic analyses were conducted without subsidies. Accordingly, hypothetical "benchmarks" (similarly treated comparisons of the economics of the conventional energy sources expressed as levelised unit costs that are likely to be displaced by renewable energy technologies) were developed. When a technology is described as "Economic", the comparison with these levelised unit costs of conventional energy sources is implied. That is, a technology could in practice be cost-effective (benefits outweigh costs), such as in certain high energy cost regions, but might not be categorised as "Economic" if it could not effectively compete with the hypothetical benchmark. This methodology and the parameters used to develop the benchmarks are fully described in Annex II.

It should be noted that there can be considerable differences in the assumptions which are used in performing economic analyses of energy supply options. Understandably, the results of independent economic analyses can vary quite substantially. Individual governments will therefore benefit from performing similar analyses for themselves, using the assumptions most appropriate to their particular energy supply situation. It should also be observed that there are differences in the methodologies used. This adds confusion to attempts to compare results from independent analyses. Finally, there has been very little accomplished on quantifying such effects as the benefits of locally-produced energy, environmental impacts, or the value of security of supply. Governments might benefit greatly from an exchange of information on economic analysis methodologies and assumptions used in these analyses and the quantification of externalities.

CONCLUSIONS TO CHAPTER II

Renewable energy sources in IEA countries constitute a substantial potential contribution to these countries' total energy requirements. Availability of the renewable energy resource and accurate resource assessment are critical factors for the success of any commercial renewable energy project. Most IEA countries have performed at least national resource assessments; many have done more detailed region or site-specific assessments as well. Two "Economic" renewable energy technologies whose expanded contribution depends most critically on site-specific resource assessments are small- to medium-size wind systems and geothermal electricity-generation and direct-heat applications.

In the last twelve years substantial progress has been made in moving every renewable energy technology studied here into a higher level of technical development. Some have reached the "Economic" stage through these efforts; though their markets may be quite limited, e.g. only occurring where resources or cost conditions are optimum. Introduction to these small markets has been important to test and further develop these technologies to meet users' requirements. At their present stage of development, many of these technologies are even more appropriate to developing countries than to IEA Member countries. Industries developed for internal renewable resource development can improve their competitiveness by expanding production to provide components and systems to these other countries.

Because market penetration has not been as widespread or as rapid as possible for "Economic" renewable energy technologies, it is useful to identify which factors are most crucial to increase market penetration. Generally, all of those that have been commercialised could substantially increase their market penetration rates by lowering production costs. Chapter III identifies institutional barriers to energy markets, and Chapter-IV discusses a variety of means which governments can use to reduce some of these problems in order to accelerate penetration of commercialized technologies.

The size at which an installation becomes economically viable is largely a function of the load and available resource at a site and of the ability to transport the energy produced to users. Technologies which produce heat directly to an end-user, such as a residence, must be matched to the size of the end-use. These types of technologies are subject to information dissemination and diffusion problems similar to those encountered by conservation technologies. These problems can include such aspects as proper installation and system integration with existing systems. The renewable energy technologies most susceptible to this problem and whose development is primarily limited by it are: active solar space and water heating, passive solar heating, daylighting, biomass (heat and anaerobic digestion) and geothermal direct heat. (Geothermal direct heat is also limited geographically.)

Mass production is another factor which could substantially lower the cost of some technologies and thereby positively influence market penetration. The renewable energy technologies most dependent on mass production, at present, are active solar space and water heating systems, small- to medium-size wind systems, parabolic troughs for industrial process heat and electricity generation, and small PV systems in remote locations. Economies of scale and capital availability for very large projects become critical for other technologies. Tidal power is the most notable technology whose very large economic size may require very innovative financing arrangements in order to expand the number of commercial projects.

Finally, negative environmental impacts can be a constraint on development of an economic technology, and, while most of the renewable energy technologies are relatively benign environmentally (i.e. environmental impacts are minor and controllable), biomass combustion technologies can produce environmental impacts to air and water. In addition, efforts to increase biomass availability (collection techniques and energy cropping) raise concerns in land-use, agriculture, and forestry policy areas.

Many others of these technologies are in early developmental stages, and will require extensive RD&D before they can be economically competitive with conventional energy systems, especially with present electricity grid networks. For the renewable energy technologies which are not in the "Economic" category, RD&D successes will be the critical factor in advancing the technology toward commercialization. Of course, RD&D can help to extend the application of technologies in the "Economic" category, but it is not critical. RD&D activities for each technology which are presently considered most important for each technology's advancement, as summarised from Annex I, are presented in Chapter V.

Because RD&D funds are becoming scarcer, it is important that these funds be used to maximum advantage. Attention to the stage of development a technology has reached (i.e. its technical status, its outlook for further development and its market attractiveness) should help ensure that the timing, scale and placement of funds are optimum.

The energy production capability from renewable energy sources and technologies is site-specific and generalisation of their technical, economic and commercial prospects is difficult and sometimes misleading. The economic analysis methodology used in this study has the benefit that it allows development of a levelised present-value unit cost for any energy supply (renewable or conventional). This in turn allows direct economic comparison between any two or more energy supplies. Because of the variations in cost components in each country and even for each renewable energy project, governments would benefit from performing their own analyses of renewable energy technology economics.

The long-term outlook for renewable energy technologies hinges, for those that are less well-developed, on technological advances and for all of them, on increases in the real costs of fossil fuels, and, of course, the timing of both is difficult to predict. Ultimately, the success of renewable energy projects will depend on whether they can compete with other energy options, such as electricity from nuclear power or coal-fired power plants or heat from coal, gas and oil. Nevertheless, for use in remote areas or in some developing countries, renewable technologies may be the only solution or one of the most attractive alternatives or, for some countries, the most acceptable in view of environmental constraints.

IEA countries have worked towards developing renewable energy technologies which seemed promising a decade ago. The results have been impressive. The prospects for and problems to be solved in each renewable energy technology are much better known. Although more work needs to be done to make renewable energy technologies important energy supply options, efforts should be concentrated on the areas which are most promising in terms of technological prospect, resource availability, environmental requirements, economic prospects, and other policy considerations.

CHAPTER III

INSTITUTIONAL AND OTHER FACTORS AFFECTING THE DEVELOPMENT OF RENEWABLE ENERGY SOURCES

Once a new technology reaches the commercial stage and its economic competitiveness is approaching that of other technologies already in the market, then institutional and other factors take on major importance as they dictate when and how the new technology will become competitive. In order to permit developing renewable energy sources to become an option for choice by the energy market and therefore begin to make their economic contribution in the energy mix of IEA countries, it is important that inadvertent institutional or other barriers which affect the development of a market be reduced as much as possible and that institutional arrangements which affect the energy market should, as far as possible, treat renewable and other sources of energy equally. This would not preclude government measures to promote or discourage a particular form of energy; rather, such measures would be introduced and controlled as a result of an explicit energy policy.

The purpose of the present chapter is to examine these factors in IEA countries and to identify ways of overcoming constraints to introducing these technologies into the market-place, which would result in increased use of renewable energy sources on an economic basis. The energy situation, of course, varies greatly among countries, but there are a number of issues which arise often enough (e.g. legal, regulatory, environmental, financial, market, and societal) to justify consideration at an international level, even though the solutions may need to be formulated on a case-by-case basis. The separate question of the ways in which governments may remove impediments and encourage the introduction of renewable energy sources is discussed in Chapter IV.

FACTORS AFFECTING THE DEVELOPMENT

Through formal or unstated energy policies, and in accordance with agreed IEA objectives, all Member countries of the IEA rely on a combination of market forces and government action to affect their energy balances. Some of these policies are directed at removing or influencing market limitations which exist in nearly all IEA countries and which are considered to be counter to the countries' overall policy goals. For example, in many cases, some or all energy prices may be judged to give inappropriate signals to consumers and investors. These price signals can have a significant impact on the development of renewable energy sources if they artificially promote those fuels which compete either directly with renewable energy sources or with electricity, which is an important market for many renewable energy sources. This particular effect is often the result of energy prices which are influenced much more strongly by short-term market considerations and investment decisions than by the government's long-term energy policy outlook. This is important for many renewable energy technologies which require lengthy development before a commercial product is available. Price signals may also be distorted where some sectors of energy industries have tended towards oligopoly or monopoly. In these cases there are severe practical problems in determining the economics of energy supply or what economic energy prices are.

Other factors besides prices influence decisions about energy supply. In fact adequate information about alternative choices in energy supply and demand management is essential and yet often lacking. Even where this information is available, however, new technologies and new products need a long time to earn users' confidence especially when they are supplied by industries which have not been well established.

These sorts of limitations in the energy market may lead governments to intervene for policy reasons. The possible impact on the development of renewable energy sources on an economic basis is one of the factors to be taken into account in developing policies in the energy supply area. The following sections discuss those factors which tend to affect the introduction of renewable energy techologies into the market.

Investment Decisions and Project Financing

To a typical investor, investments in renewable energy technologies may appear quite different from those in conventional energy projects. For example, most conventional energy sources have established infrastructures for which costs are "sunk" or financed over a very long time, often with government assistance. Investments in renewable energy technologies often include the establishment of infrastructure for the technology which can add significantly to its capital cost.

The supply of conventional sources of energy is, in most IEA countries, in the hands of large utilities or other bodies for which energy supply is a central activity while the supply of renewable energy sources is more disaggregated. In some cases, large companies, such as electric utilities, may themselves supply or promote the supply of renewable energy sources as a part of their activities. In other cases, renewable energy sources are a form of self-supply outside the energy sector and may involve only private consumers. Firms tend to give priority in the use of inevitably limited financial resources to investments which will promote their main activity and to accept a lower rate of return on such investments than on peripheral investments. This gives rise to situations in which, in practice, a lower rate of return is accepted for an investment in a conventional source of energy than for renewable energy sources. For example, whereas electric utilities generally calculate the rate of return on investment over the investment's life period, which is often much longer than ten years, manufacturing companies and private consumers tend to seek payback periods of one or two years which penalises capital-intensive, low-operating cost technologies such as most renewable energy technologies.

In some Member countries of the IEA, some or all of the main energy supply industries are nationalised. Nationalised industries are normally required to judge investment proposals against a public sector discount rate which, in IEA countries, varies between 5% and 10% in real terms and, in most cases, is at, or close to, the lower figure. Although this discount rate is intended to reflect the rate of return on capital in the economy as a whole, in practice it may tend to be lower than the discount rate accepted by the private sector, particularly in investment on peripheral activities. Therefore, in these countries, governments may obtain a higher benefit from renewable energy technology development than would the private sector.

Decisions on the level of acceptable rate of return for private investors and access to capital are directly affected by the degree of perceived risk of the investments. Very high rates of return, as much as 50% return on investment or less than one year payback, are typical for new small business ventures. Perceived risk is a direct function of the size of the company and its asset base and the track record of the technology, i.e. the information available on the performance and reliability of the technology. Large multinational corporations, regulated utilities, and nationalised industries consequently have easier access to capital on better terms. To the degree that imperfect information is available on an investment (or corroborating information is not available) and the promotors of renewables are small companies, the rate of return must be higher, and therefore fewer investments will be economic. This fact underscores the need for ensuring that more and better information on the performance and reliability of renewable energy sources is available to investors. It also points out a need for investment capital at reasonable rates for renewable energy projects and for private sector firms developing these new technologies.

Governments have only a limited ability to influence the tendency of firms to accept a lower rate of return on their main, rather than on their peripheral, activities or of investors to require higher return to minimise risk. In cases where it is clear that such a tendency prevents adequate investment in renewable energy sources, a degree of government support for those energies may be justified in the interests of the government's perception of optimal resource allocation. The investment criteria used by nationalised industries are more directly within the control of governments, and it is important that these criteria should be set in a way which does not introduce an unintentional bias against investment in renewable energies.

It has been noted that if renewable energies are to achieve their potential in the energy mix, financing for projects needs to be available, and yet the high financial risk associated with relatively new and limited distribution technologies severely limits access to capital. Some renewable energy projects have proceeded via access to capital available on normal commercial terms by taking advantage of innovative financing arrangements. The possibility of such arrangements varies with the tax system in different countries. In the United States, new markets were opened for renewable energy sources as a result of innovative financing. For example, companies were created to specialise in ten-year financing of residential solar systems. Venture capital investment packages were developed which allowed rapid wind farm development. These examples show the capacity of the financial community to develop the means to finance "risky" projects when interest rates, price signals and incentives can be combined to provide attractive investment packages.

Information on Renewable Energy Resources

There is lack of dissemination of information about availability of renewable energy resources on a local level. While an inventory of renewable resources is one of the pre-requisites to producing reliable assessments of the contribution potential and for devising sound programmes for renewable technology development and market introduction on a national level, individual projects also need site-specific resource assessments in order to obtain financing. The accuracy and adequacy of characterisation in basic resource data (e.g. insolation, wind speed) needs improvement to increase the deployment of renewable energy technologies in many countries. This applies especially to developing countries and to remote regions. For optimal system design and thus for adequate investment calculations, knowledge of the magnitude and pattern of local availability of renewable energy sources together with systems performance analysis is essential, which for some renewable energy technologies involves long-term and detailed observations.

The Role of the Electric Utilities

While both gas and electric utilities are involved in development of some renewable energy technologies, electric utilities play a key role in the market penetration of those technologies which produce electricity because they often purchase surplus electricity from renewable energy producers. At the same time, renewable energy producers may depend on the utilities to supply them with electricity when supply from their own sources is interrupted. There are, in addition, many instances of utilities supporting the research and development of renewable energy technologies. This support most often takes the form of demonstration projects, as in the cases of the national electricity board (ENEL) in Italy (wind and solar), Unidad Electrica S.A. (UNESA) in Spain (solar energy), the Central Electricity Generating Board (wind, wave and tidal power) and the North of Scotland Hydro Electric Board (wind and wave power) in the United Kingdom and the Tennessee Valley Authority (solar water heating programme) in the United States. In some cases, utilities use renewable energies either to generate electricity, such as in Spain, Sweden, the United Kingdom and the United States, or for combined *heat* and electricity production plants such as in Finland and Sweden.

General principles for the purchase of electricity by utilities from auto-producers (or self-generators) were suggested in the Study "Electricity in IEA Countries - Issues and Outlook" published in 1985*.

These principles also apply to the purchase of electricity produced by renewable energy producers. If utilities are to make a sound choice between producing electricity themselves or buying it from others, the purchase price ought to reflect the reduction in the costs of the utility resulting from buying the power, i.e. the utility's avoided costs, composed of the reduction in fuel costs, other operating costs and, if the purchase enables the utility to avoid provision of new capacity, a capacity element. The value of purchased power to a utility varies with time. Power supplied from renewable energies is sometimes not firmly guaranteed for peak periods; therefore determining the amount of the capacity element to be included in the price paid by the utility will often be complicated. Similarly, the price charged by the utilities for back-up supplies to those who depend on renewable energies ought to reflect the cost of making those supplies available.

In some cases, electric utilities may have departed from these principles and have used their special position as monopolies to discourage renewable energies, either by offering an unduly low price for the electricity they purchase, or by imposing onerous terms for back-up supplies. In the Netherlands, the price paid to independent wind power producers is slightly less than the cost of the fuel saved. In the United Kingdom, in the past the private supplier of electricity had to pay significant standby charges

---

* Electricity in IEA Countries - Issues and Outlook; OECD/IEA 1985

imposed by local electricity boards which rendered small wind turbine machines uneconomical if they depended on grid connections. However, this problem is now very much reduced as utilities gain experience with private generation.

In other cases, where there has been a desire to promote the development of alternative energy sources, policies have been developed, such as in the United States under the Public Utility Regulatory Policies Act of 1978, to require utilities to purchase, at their own avoided costs, the co-generated or renewable-energy-source power that is made available. These policies were considered necessary to enable renewable energy sources to compete on a more equal basis with utilities which enjoy a regulated rate of return, monopoly of status and, generally, much better access to capital. In a number of States, particularly California, this Act may have been operated in such a way that utilities have been required to pay more than their costs. Part of the rapid development of renewable energies in California is due to this factor.

In a long-term perspective, most electric utilities will gain from stimulating the development of renewable energy use. In most countries electricity demand grows faster than energy demand in general. At the same time, new large electricity generation plants are becoming more and more costly, and the procedure of obtaining approval for new plants is more complex and difficult. Thus utilities will face a situation where capacity is scarce and expensive. In these circumstances it may be economical for electric utilities to stimulate the use of renewable energy, especially for use during peak times, and to pay not only for avoided energy costs but also for avoided capacity costs. This will tend to make more renewable energy sources economic. Using the marginal (avoided) cost concept in calculating a levelised present value cost of all capacity supply options should indicate when it is optimum for a utility to begin to invest in the renewable energy sources available to it.

Legal and Regulatory Factors

The production of renewable energies, like other forms of energy, is necessarily subject to a legal and regulatory framework dealing with such issues as safety and the protection of the environment. For the reasons discussed above, renewable energies are also affected indirectly by the regulatory framework for electricity tariffs. Much of the legal framework under which the energy industries operate was introduced before the advent of renewable energy technologies. Nevertheless, it has often been found possible to deal with their position within the present legal framework.

In the United States, various legal and statutory methods have been adopted to protect rights to sunlight. A number of States have enacted a statutory declaration that it is in the public interest that solar energy appliances be encouraged (e.g. in the California Civil Code). In Arizona, Colorado, California and Maryland, legislation declares void and unenforceable covenants which impede or expressly prohibit the installation of solar devices. The California Solar Shade Control Act of 1978, included in the California Public Resources Code, gives wide protection against blocking of an established solar access by the planting of trees and shrubs, using government's public nuisance authorities. Protection of solar access has been developed by the legislation of New Mexico and Wyoming, which declares that the right to use solar energy is a property right. A solar user who uses sunlight to collect solar energy is able to protect his property rights to the sunlight by preventing any development from occurring on neighbouring property unless he is compensated under the transferability clause by the purchase of his right.

To encourage the use of wind energy, the California Energy Commission in 1982 drafted a "Model Ordinance for Small Wind Energy Conversion Systems" designed for adoption by the various local government authorities. One provision of this model ordinance safeguards sufficient wind access to ensure the efficient operation of the wind generator.

In a few cases, the application of an existing legal framework can inadvertently act to hinder the development of renewable applications. In Australia, it is estimated that an important legal barrier to the use of solar devices is restrictive covenants which are not aimed directly at solar devices but have the effect of excluding any such installations which are visible to neighbours. For example, covenants which preclude "appliances and installations on roofs", "the construction of a roof containing any reflective material", or "exterior changes or additions" make it difficult, if not impossible, for the covenantor to install a solar device. To overcome these constraints, some Australian States, notably South Australia and New South Wales, are planning legislation for solar access.

The question of the effect of standards and codes on the efficiency and reliability of renewable energy devices favouring the use of renewable applications by consumers is discussed in Chapter IV.

Legal positions will, of course, vary from country to country, and generalisations are difficult. Legal and regulatory frameworks do not appear to impede the development of renewable energies significantly. Environmental controls, for example, are no more severe than those applied to other sources of energy. There are, however, a number of areas which, in some Member countries, may require attention. In addition to the need to ensure that utilities operate in a way which optimises resource allocation as discussed above, they include:

—   The development of geothermal energy sources often faces complex and slow regulatory procedures. Rigid standards on the use of underground waters can create difficulties in obtaining permits to produce geothermal fluids in commercial quantities.

—   Safety concerns are largely a function of the quality of construction and the manner in which the installation is operated. The development of at least nationally agreed procedures for safety evaluation could aid the development of the renewable energy sources industry, while at the same time strengthening confidence among potential users. In the United States, solar installation codes and certification procedures have been adopted by individual jurisdictions and, in some areas, on a national basis.

— Aesthetic questions enter into considerations of the location of renewable energy systems. For example, the introduction of a single or several wind machines has visual effects similar to power transmission pylons on the landscape. A seashore windfarm site is much more likely to generate strong public opposition than would a remote site. In procedures authorising the installation of renewable devices, the problem of aesthetics has been taken into consideration in many cases. Belgium reports that potential visual pollution is taken into account when processing a building licence for a Wind Energy Conversion System. In the United Kingdom, aesthetic considerations may force the siting of the large wind farms offshore. In the United States, for windfarms, an Environmental Impact Assessment (which includes assessment of visual impacts) is sometimes required. The introduction of passive solar technologies in architectural design can affect building aesthetics. However, such new architectural designs are well accepted by potential owners of passive buildings where comfort or function is improved. Due to the subjective nature of the matter, aesthetic problems have to be resolved on a case-by-case basis.

Industrial Factors

Given the stage of development of most renewable energy technologies, as discussed in Chapter II, it is not surprising that the development of industries to provide the variety of equipment needed for renewable energy technologies is at an early stage; although, progress has been made in a number of countries where some renewable energies are more developed. Infrastructure similar to that involved in the resource exploration, production, conversion and distribution or transmission of conventional energy resources, but dedicated to renewable energy, is developing. In the absence of a dedicated infrastructure, reliance must be placed on existing infrastructure which tends to treat renewable energy sources as a sideline opportunity (e.g. plumbing industry for solar domestic water heaters, electric utilities for wind) with only limited success. This approach can be useful during the initial stages of market penetration but is not adequate for full penetration.

Co-ordination of Various Interest Groups

As with conventional energy projects, the building of large renewable energy projects in areas close to populations generally involves various socio-economic interests. Unlike conventional projects, renewable energy projects have many unknown or uncertain impacts causing concern, while at the same time they often have considerable support because of their generally benign characteristics. Good co-ordination of all these interests is indispensable to allow for a wide acceptability of renewable energy sources by different local interest groups.

An example of the types of interests represented is provided by the case of the preliminary study on the feasibility of a tidal power barrage on the Severn Estuary (United Kingdom). The developers took into consideration a wide range of interests involved in the construction of such a barrage. These concerned mainly the opportunities in the area for recreation and boating, the possible damages to salmon fisheries, the socio-economic advantages of a public road across the estuary carried by the barrage, the problems of accessibility of the ports in the Severn Estuary, the importance of a barrage for local employment and the future development of agriculture and industry around the estuary. A careful investigation into the relative importance of all these local interest groups was carried out by the Severn Barrage Committee and led to a balanced solution between the various interests recommending that attention should be concentrated on the Inner Barrage. This work has now been taken much further by the Severn Tidal Power Group. A full regional study will be required before the work begins. Likewise in France, the La Rance tidal barrage was designed after in-depth consultations with the groups of interests representative of the different categories of users such as fishing, leisure boating, fish-breeding, oyster breeding, ship-building and maintenance thus permitting prompt acceptability of the tidal barrage.

To ensure that renewable energy projects are not held up by concerns of different interest groups, the government agencies in charge of protecting these varied interests should be made aware of each renewable energy project and its importance as early as possible in project development. Once informed, these agencies can then act to avoid unnecessary delays and to assist in developing acceptable solutions to problems which the energy projects create.

CONCLUSIONS TO CHAPTER III

Institutional and other related factors are extremely important if economic renewable energies are to become well commercialized. There are, on the one hand, a number of inadvertent but widespread biases that may prevent renewable energies from reaching their full market potential and, on the other hand, there are few instances of inadvertent bias in favour of renewable energies. Government energy policy is directly involved with influencing these factors and biases. It is important that the institutional arrangements should treat renewable and other sources of energy as equally as possible. Numerous limitations of the operation of capital markets and in investors' decision making processes make it hard for renewable energy projects to compete with conventional energy projects, from the standpoint of obtaining the low-cost financing to build projects. The role of capital and financing institutions will thus be critical both for project finance and for financing renewable energy companies. Accurate and unbiased information on the performance of renewable energy technologies as well as on the adequacy of the renewable energy resource at proposed project sites will help to reduce perceived investment risk.

Electric utilities (and their regulators) play a key role in market penetration of these technologies because so many produce electricity and rely on these utilities to purchase the power generated and provide back-up power in the case of intermittent sources. Fair purchase prices can be developed to reflect the costs and benefits to the utility as with any purchased power. Electric utilities themselves will undoubtedly continue to develop their own renewable energy projects where the resource is available and as capacity costs and requirements grow.

Legal and regulatory factors tend to impinge to some degree on any renewable energy project in all countries. While these factors are most appropriately identified and minimised on a factor-by-factor and technolo-gy-by-technology basis, it is important to recognize that such barriers will occur and to identify and reduce them early in a technology's market introduction phase to avoid unnecessary delays.

The situation varies among countries, but in general the infrastructure and institutional needs common to any energy technology are relatively undeveloped for renewable energies. For those renewable energy technolo-gies that have already moved into the "Economic" stage, these needs have been met; not easily or without problems, but by and large, they evolved with the technology's industry and with government assistance. While it is important not to confuse normal market resistance to new technologies with barriers that may appear to be peculiar to renewable energy, the geographically-specific nature of many renewable energy projects makes it important that local groups are aware of and involved in the planning and co-ordination of the project with other local interests again to avoid delays and opposition.

CHAPTER IV

GOVERNMENT SUPPORT FOR RENEWABLE ENERGIES

INTRODUCTION

Experience indicates that the development of new sources of energy requires long lead times. For each new source the energy supplying capabilities must first be recognised, then the technologies developed and tested to transform the energy into usable forms, and finally the technologies must be made available to the user (through the development of infrastructure, e.g. utility, industry, building). Potential users must decide that it is to their advantage to use the new source and its attendant technology in view of their former energy supply and technology. For example, oil and natural gas were introduced into the energy market over a long period after their discovery even though these fuels were attractive from the beginning due to their cost advantages, flexibility of use and convenience in transportation, storage and combustion. The technologies for facilitating their use were developed, primarily by market mechanisms, assisted by government support in many countries, such as in support of the development of pipelines or in tax advantages.

Governments become involved in influencing energy supply development for a variety of reasons related to national interests. Many IEA Member governments are counting on substantial additional contributions from renewable energies to their energy supplies as early as the year 2000. Renewable energy sources appear attractive to governments because of their potential contribution to the goals of energy diversification and security of supply. In addition, the increased importance of environmental considerations in overall energy policies has encouraged increased use of renewable energy sources since they are in most cases more environmentally benign than conventional energy sources. The long lead times required for R&D and large capital investments required before the technologies could be proven to be economic have meant that the industries that would ultimately benefit have not always been prepared or able to undertake the necessary R&D at their own risk and expense. This has brought governments to decide to play a central role in the development of renewable energy technologies, with the expectation that the competitiveness of renewable energy sources with conventional energy sources would result.

In the past ten years, many technical and economic uncertainties related to the development of renewable energies have been addressed. As a result, governments can now expect industries to play a greater role in the future development of many of "Economic" renewable energy technologies. Based on the experience of the last decade, it is now useful to attempt to assess the different types of government support, to find out which government policies and actions have played an important role in reaching the "Economic" stage, and to apply these lessons to those renewable energy technologies still under development.

All IEA Member countries have instituted policies and programmes aimed at encouraging increased use of renewable energies that can contribute to desired national energy supply balances. Government action is not limited to RD&D funding but includes a variety of policies and activities necessary to secure the economic and commercial environment to facilitate the introduction of renewable energy technologies when ready. The specific policies reflect national energy situations and interests, both in terms of R&D areas, and financial, fiscal and other support measures. Government activities fall into three general categories:

—   direct funding of research, development and demonstration projects;

—   development of institutional frameworks to facilitate the use of renewable energies (e.g. legislation, regulation); and

—   measures to improve economic competitiveness and market penetration of renewable energy technologies (e.g. incentives, information dissemination).

The timing and appropriateness of these various government activities depends upon the stage of development (as described in Chapter II) of each of the renewable energy technologies. In general, to be effective, government activities will differ greatly as technologies advance from the research and development phase, through the pre-commercialization phase, into the final full-scale commercialization phase. RD&D activities are most appropriate for technologies which are in the "Under-Development" and "Future-Technology" stages, especially since in these stages it is less likely that industry would conduct the work in the absence of government activity. The second and third categories of activity, institutional facilitation and marketing facilitation, are only appropriate for technologies which have reached the "Economic" or "Commercial-with-Incentives" stages. If specific government activities are initiated or terminated prematurely, or continued too long, the results could be more prejudicial to the sound development than if no government activity had ever been conducted. Government support, in its many forms, is most effective when there is careful planning, priority setting and on-going evaluation.

DIRECT FUNDING OF RESEARCH DEVELOPMENT
AND DEMONSTRATION

Before 1973-74 renewable energy RD&D activities had been carried out in only a few countries and primarily in areas of basic research. For example, some government-funded efforts had been made in the field of photovoltaic conversion for space applications. In some countries, which had more favourable conditions for solar energy utilisation (Australia, the United States and France), successful work had been going on since the 1940s, with some substantive support from the public sector.

During the rapid oil price increases between 1973 and 1980, the total budgets of IEA Member countries for renewable energy RD&D increased dramatically. R&D expenditures then declined to level out at about the 1978 level of effort. The decline is due to major R&D expenditure cuts in the United States. Without the effect of United States spending the R&D budget steadily increased until 1981 where expenditures have more or less levelled off, although 1985 showed a marked decrease again. Table 3 illustrates their evolution over the past nine years. For renewable energy R&D activities, the first four or five years were a definition phase: to develop analytical data

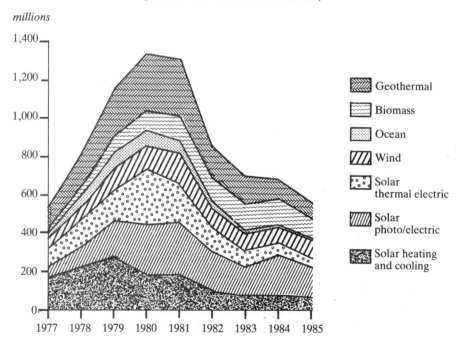

FIGURE 1
IEA GOVERNMENT RD&D EXPENDITURES FOR RENEWABLE ENERGIES
(Millions 1985 United States Dollars)

Source: Energy Policies and Programmes of the IEA Countries, 1985 Review; OECD/IEA, Paris 1986

bases and methodologies, to implement research programmes, and to develop and demonstrate components and systems. These activities have led, in the last few years, towards greater selectivity, based on earlier RD&D results and experience with bringing some technologies into commercial and economic stages, and have also showed continuing concentration of resources on fewer projects through reduction of technology areas receiving priority consideration.

Figures 1 and 2 provide a breakdown of RD&D expenditures by technology. After 1980 these figures reflect tightened government budgets and changing national policies and perceptions regarding the state of development of the various renewable energy technologies and their national energy contribution or export potentials. In general, RD&D funding in relative terms for photovoltaics and biomass has increased, whereas that for ocean energy and solar thermal has decreased, and that for active and passive solar, wind, and geothermal energy has remained relatively unchanged. While each country considers many factors in establishing its specific RD&D funding distribution, it is generally believed that the above changes are primarily because of either successful cost reduction for some technologies, revised estimates of potential energy contributions or estimated high costs associated with further development for other technologies.

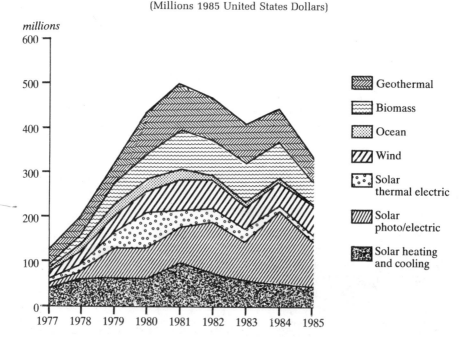

FIGURE 2
IEA GOVERNMENT RD&D EXPENDITURES FOR RENEWABLE ENERGIES, EXCLUDING UNITED STATES
(Millions 1985 United States Dollars)

Source: Energy Policies and Programmes of the IEA Countries, 1985 Review; OECD/IEA, Paris 1986

The cumulative government funding for renewable energy RD&D between 1977 and 1985 amounted to over US $7 billion (1985 prices) for IEA countries. In 1985, IEA countries allocated about 8% of their aggregate R&D energy budgets for renewable technologies. Indeed, recent total IEA Members' public expenditure on renewable energy research and development was in the range of that devoted to coal technology development.

After the steep decline since the peak of 1980, Table 3 shows some levelling off in 1984, but still with a slight downward trend. The proportion of energy RD&D budgets devoted to renewable energy sources by many of the countries with smaller economies is substantially above the overall IEA average. This tends to reflect a more emphatic commitment to indigenous resource development to the extent that the cost of the RD&D can be handled within national budgetary constraints. The diversity of national choices among the various renewable energy technologies is illustrated in Table 4, which shows the nine countries with individual budgets for renewable energy in excess of an equivalent of US $15 million. These nine countries' expenditures account for nearly 90% of all government expenditures on renewable energy. It should be noted that industry expenditures for R&D are not reported in any of the previous tables. These expenditures have been substantial; although it is not possible to report an amount because these figures are not available.

Almost all IEA countries maintain at least minimum programmes (e.g. research progress monitoring) in most renewable energy technology areas. This approach is related to a desire to keep abreast of developments which take place in the various technologies. In the majority of countries there is also a clear interest in one or two technologies which are considered to be particularly suitable for use as an indigenous resource. Other countries wish to achieve a leadership position in one or two technologies with the possibility of stimulating exports. In the case of the United States, however, important activities are being conducted across the complete spectrum. Canada has emphasized biomass and solar heating (although expenditure levels have been significantly reduced); Germany and Italy expect to develop an export business in photovoltaic systems (in addition to using such systems domestically); the Netherlands and Denmark have a special interest in wind; Sweden's interest is in biomass and to a lesser extent in solar heating; and the United Kingdom is pursuing the more economically promising options including biomass, passive solar design, hydro, wind, geothermal hot dry rock and tidal energy.

Apart from the activities of its individual Member States, the Commission of the European Communities also carries out and finances a considerable amount of R&D and Demonstration work. In addition to research undertaken at its own centre, the Commission issues calls for proposals for both RD&D projects, covering most of the renewable energy forms. On the R&D side, between the beginning of the programme in 1975 and 1985 over 1 400 projects have been allocated finance totalling 175 Mecu[1], and

---

1. Mecu = Million European Units of Account. Over the period 1975 to 1985 1 ecu was worth on average about US $0.95.

TABLE 3

IEA GOVERNMENT RD&D BUDGETS FOR RENEWABLE ENERGY IN 1985 UNITED STATES DOLLARS
(Millions)

	1977	1978	1979	1980	1981	1982	1983	1984	1985
Canada	13.4	32.0	35.6	36.9	57.3	47.9	52.5	41.2	23.5
United States	394.8	591.9	801.8	876.2	787.1	378.0	284.6	232.8	220.3
Japan	23.3	29.1	37.5	91.3	89.8	93.3	84.7	78.4	70.5
Australia	5.0	n.a.	7.6	10.7	13.1	n.a.	10.3	n.a.	n.a.
New Zealand	1.5	2.6	2.3	3.1	3.1	2.3	2.6	2.4	2.0
Austria	3.0	3.9	4.5	5.5	4.2	4.0	4.4	2.3	2.1
Belgium	1.2	1.4	3.7	6.2	9.7	4.8	6.9	7.9	7.7
Denmark	2.4	4.6	8.7	4.1	2.6	2.7	2.5	2.5	2.2
Germany	13.5	22.9	50.1	52.5	55.2	83.8	40.3	49.1	42.9
Greece	0.4	0.5	1.7	6.5	8.0	2.1	2.0	2.9	4.4
Ireland	0.4	0.6	1.1	1.2	3.9	3.3	1.7	0.7	0.5
Italy	3.7	8.0	7.8	14.9	28.2	13.9	24.2	48.5	17.0
Netherlands	9.5	10.7	12.4	13.9	16.6	16.1	18.7	15.1	33.2
Norway	0.2	2.2	4.7	4.3	3.5	2.8	2.9	2.4	2.2
Portugal	0	0	0	0.6	0.6	0.8	1.1	1.6	n.a.
Spain	4.8	3.2	7.7	19.3	14.1	14.1	30.1	42.5	15.7
Sweden	10.5	20.1	39.9	33.9	43.1	41.3	30.3	24.2	19.3
Switzerland	3.0	4.8	7.3	7.3	7.7	6.5	7.4	6.9	5.9
Turkey	0	0	0	0.4	0.3	0.3	0.5	0.5	0.4
United Kingdom	4.1	11.7	20.5	18.2	26.3	19.7	15.5	17.3	15.7
IEA Total	494.8	750.4	1 054.8	1 207.0	1 174.6	737.6	623.5	579.3	485.4

Source: *Energy Policies and Programmes of the IEA Countries, 1985 Review*, OECD, Paris, 1986

TABLE 4
BREAKDOWN OF MAJOR GOVERNMENT FUNDED RD&D IN 1985
RENEWABLE ENERGY RD&D ACTIVITIES IN 1985 UNITED STATES DOLLARS
(Millions)

	Solar Heating	Solar Photo Electric	Solar Thermal Electric	Wind	Ocean	Biomass	Geo-thermal	Total	% of National Budget
Canada	5.93	3.29	0	4.54	0.22	9.08	0.44	23.50	6.24
Germany	7.64	18.10	6.11	6.52	0	0.20	4.28	42.85	7.81
Italy	4.59	6.04	0.06	3.77	0	2.23	0.30	17.00	2.99
Japan	2.49	35.19	0.83	1.25	1.43	7.69	21.62	70.51	4.53
Netherlands[1]	0	4.18	0	22.06	0	2.71	4.21	33.16	28.89
Spain[1,2]	0	23.52	0	11.47	0.06	3.00	1.18	39.22	26.56
Sweden	4.37	0.51	0.07	1.86	0.12	11.87	0.46	19.26	23.40
United Kingdom	1.15	0	0.13	8.47	0.51	0.64	4.75	15.66	4.30
United States	12.90	56.90	35.80	29.50	4.70	49.70	30.80	220.30	9.76
IEA Total	39.07	147.73	43.00	89.44	7.04	87.12	68.04	481.46	

1. The figures for the Netherlands and Spain apply for the groups Solar Heating, Solar Photo Electric and Solar Thermal Electric combined.
2. Data is from 1984 submission.

Source: *Energy Policies and Programmes of the IEA Countries, 1985 Review*, OECD, Paris, 1986

530 renewables demonstration projects have been awarded support totalling 167 Mecu since the Demonstration progamme began in 1978. New programmes have recently been adopted in both fields, with budgets averaging about 45 Mecu/year for R&D and 90 Mecu/year for Demonstration (although this latter figure includes other sectors such as Energy Saving). The Commission is also giving increasing emphasis to more general measures for facilitating the economic emergence of the renewable energies throughout the Community.

INSTITUTIONAL FRAMEWORK TO FACILITATE THE USE OF RENEWABLE ENERGY

Government can play a role in encouraging renewable energy technologies once they have achieved at least the demonstration or the commercial stage, through such means as data bases, standards and codes for renewable energy devices, public information, leadership activities in public institutions, and training. The following sections provide examples of these activities.

Data Bases

A number of governments have recognised the need for adequate resource information (or resource "prospecting") as a precursor to project design and site selection. Annex I provides several examples of government-sponsored resource asessments. Wind atlases are a specific example already cited earlier. Publication of solar insolation data is another government activity which can assist private developers in site selection and in obtaining financing. Equipment loan programmes have also been successful in assisting resource prospecting. Annex I identifies additional areas where resource assessment may be critical to further development of a resource such as reservoir definition for geothermal developments.

Standards, Codes and Equipment Testing

Almost all IEA governments or other interested institutions already have developed technical standards and codes on the efficiency and reliability of commercially available renewable devices notably in the technologies of solar systems, wind energy conversion systems, and passive solar construction. Such regulations developed for newly-introduced commercial technologies reflect the conditions of utilisation that consumers normally encounter. These reliable test procedures ensure growth of renewable energy sources and encourage innovative developments.

The development of standards and testing procedures for solar systems in several countries provides a good example. The Solar Manufacturers' Association, wishing to improve the reliability of solar water heaters, has motivated the Greek government to set up a solar collector test facility for mandatory testing and certification of collectors and solar systems on the basis of their standards adopted in 1981. In Italy, Ente Nazionale Elettricita, the national electric utility which has a central role in the promotion of solar water heating, has developed technical specifications for all solar components and a methodology for testing collector performance and durability at the Phoebus Research Centre in Southern Italy. In Japan, solar hot water systems have been standardized by the Japanese Industrial Standards which were set up by the Ministry of International Trade and Industry. In the United States, the Solar Rating and Certification Corporation (SRCC), a private industry-funded corporation, sponsors national testing, rating and certification programmes for solar collectors and solar water heaters. The Tennessee Valley Authority (TVA), in the framework of its Solar Water Heating Program, tests all solar systems before making them eligible for the programme. If a system passes the test, TVA includes the company on a list of approved suppliers from which customers must buy their equipment in order to qualify for a loan. In the United Kingdom, the British Standards Institution has published Codes of Practice for the installation of solar heating systems. The Aerospace Technology National Institute (INTA) in Spain has developed technical standards for solar components and a methodology for testing collector performance as part of a collector certification process.

Public Information and Training Activities

Information dissemination to improve credibility and public awareness is indispensable once renewable energy technologies become commercially available to encourage the market penetration of these technologies. A lack of reliable information on operation and maintenance data for a renewable system or a misunderstanding of the technical opportunities for a renewable energy development have sometimes caused negative reactions about the practical applications of renewable energy technologies. Reluctance to use renewable energy sources may be further aggravated by reports of unfavourable early experiences with some poorly designed, badly installed renewable energy technology systems.

Many technical problems in the initial stages of commercialization affecting performance, reliability, and durability of some renewable energy devices resulted from improper installation and poor maintenance by mechanics who were insufficiently trained or inexperienced in specific systems. This problem has been overcome for technologies that have achieved substantial market penetration by developing an adequate technical and practical information base, specialised training, and installer licensing requirements. For example, for the past several years the American Institute of Architects has sponsored a multi-faceted programme for the successful utilisation of passive solar techniques which has helped thousands of professionals. And in a number of IEA countries, the solar-energy-industry associations have offered installer training courses.

In Canada, twelve regional Conservation and Renewables Energy Offices have been in operation since April 1981. These were established to help deliver the Canada Oil Substitution Program (COSP) which assisted households in converting from oil to natural gas, electricity and renewable energy sources (mainly wood). These offices, which also help deliver other renewable energy programmes, have set up seminars for technicians, processed applications and provided information. In 1984-85, these offices were expected to handle approximately 22 000 enquiries per month.

Considerable efforts have already been made in all IEA countries to develop reliable and appropriate information for potential consumers through technical guides and booklets, seminars, training, mass media campaigns and special information from researchers. For commercially-available renewable technologies the information base has been growing to a point where it is now largely available to the different categories of users of renewable energy sources — technicians, installers, potential purchasers. As some of the less developed renewable energy technologies become commercially available, similar programmes for public information and awareness and installer training and certification programmes can help assure commercial success.

Leadership Activities in Public Institutions

The exemplary role of governments or other public bodies which install renewable energy systems in government office buildings and social and other facilities can stimulate the pre-commercialization of renewable energy devices by providing demonstration "showcases" for those technologies which have reached the commercialization stage.

There are many successful examples of this type of government role. Unfortunately, however, too often in the rush to demonstrate a technology, an application has been chosen that was premature (i.e. pre-commercial). For example, in Canada, the Purchase and Use of Solar Heating programme (PUSH) was established to install solar heating systems in federal government buildings, including one of the Parliament buildings in Ottawa, a fish hatchery in New Brunswick, some Canadian National Railways wagon washing facilities and the Halifax airport. PUSH spent only CDN $35.9 million over four years out of the original five-year programme budget of CDN $125 million and was terminated in June 1983. Slow start-up, administrative problems, and inappropriate programme design were the main reasons for the relatively low expenditure levels and the limited success of this programme. This type of premature promotion should be avoided because it can be damaging to more appropriate future efforts to demonstrate a technology where there are lingering doubts about why the previous effort did not work.

MEASURES TO IMPROVE ECONOMIC COMPETITIVENESS AND MARKET PENETRATION OF RENEWABLE ENERGY TECHNOLOGIES

Fiscal and financial incentives have been used by some governments to promote commercialization of renewable energy technologies by enhancing their economic competitiveness. This type of incentive is especially appropriate in the early market penetration stage where there is a gap between price affordability and national goals and where incentives would help establish these early markets. The following sections illustrate several types of incentives which have been used.

Fiscal Incentives

Exemption from specific taxes, accelerated depreciation, and tax deductions are the types of fiscal incentives most commonly used. Where it is appropriate to a country's tax structures, tax reductions and accelerated depreciation schemes can be established to reduce the effective purchase price of renewable energy equipment. Some countries have also introduced taxation on conventional energy sources, such as an oil tax. The revenue from these taxes has then been partly used to finance grant programmes for conservation and/or renewable energy technologies.

Examples of exemption from specific taxes: In early 1986, Australia applied a sales tax of 10% on all domestic water heaters other than solar. In Portugal, wind, solar and geothermal equipment manufactured with domestic or foreign designs are exempted from half of the sales tax. In Spain, the Energy Conservation Law allows reductions of 95% on import duties on any equipment related to renewable energy sources, provided that similar equipment is not domestically manufactured and that approval is obtained from the Ministry of Industry and Energy on a case-by-case basis. Around 65% of Spain's renewable energy users have applied for duty reductions.

Examples of accelerated depreciation, tax credits, and tax deductions: Belgium implemented accelerated depreciation rates for companies investing in renewable energy sources with the possibility of amortizing 135% of the investment in five years. In Canada, a three-year accelerated write-off is available for the capital cost of most renewable energy systems. In Japan, a tax credit system was established in 1981 for three years and then extended for another two-year period. This system allowed for either a 30% special depreciation or a 7% tax deduction in the first year for equipment utilising geothermal energy, solar energy and equipment utilising waste heat. When this system expires in 1986, a modified system will be in force until 1988. The new system will provide a 20% special depreciation rate.

In Germany an accelerated depreciation of 10% over ten years for private investors has been established for renewable energy.

In the United States, through complex systems of accelerated depreciation and tax deductions, the Federal and several State Governments have promoted the utilization of renewable energy technologies. United States federal tax credits offered for solar, geothermal and wind energy applications included:

— a 40% residential non-refundable tax credit available to homeowners on the first US $10 000 of their solar or wind investment which were installed in new or existing residences before 31st December 1985;

— a 15% residential credit for landlords for solar investments; and

— an additional 15% business non-refundable investment tax credit for solar investments above the usual 10% business investment credit.

Other tax credits existed for investors in biomass, small hydropower and geothermal energy. Capital investments in biomass and small-scale hydropower were made more attractive through an extra 10% business investment tax credit (total of 20%) for certain types of plant and equipment.

These business credits have been subsequently replaced by the following schedule.

	1986 %	1987 %	1988 %
Solar	15	12	10
Geothermal	15	10	10
Solar Thermal	15	15	15
Biomass	15	10	0
Small Hydro-electric	11	0	0

California introduced a 10% solar tax credit in 1976 — two years before the federal tax credit began. This credit allowed individuals and businesses that purchased renewable energy equipment for producing heat or electricity to deduct 10% of the cost, including installation, from their state taxes. In September 1977, California passed additional legislation to increase the solar tax credit to 55% with a US $3 000 maximum. The credit was reduced to 50% in 1984, and will expire at the end of 1986.

The Accelerated Cost Reduction System for figuring depreciation in the United States allowed some renewable energy technologies such as wind systems to be fully depreciated in five years. This favourable treatment was continued in the tax reform legislation of 1986.

Financial Incentives

Financial incentives include grants, subsidies, low interest and long-term loans and loan guarantees. The following examples describe some applications of grants and loans.

Examples of Grant Programmes: Commercial investors in Germany can receive a 7.5% grant without limitation for investments in new, more efficient energy systems and systems utilising renewable energy sources. An early evaluation of the programme showed that almost 90% of the incentives were used for energy-saving activities and only about 2% for renewable energy systems. At least for the German case, it can be pointed out that these programmes must be carefully designed and periodically evaluated against the objectives in order to avoid undesired or imbalanced effects. However, it should be noted that the German fiscal incentives nevertheless contributed substantial energy savings and an increased public awareness of energy conservation and renewable energy technologies.

In Canada, various grant programmes were offered, such as the Forest Industry Renewable Energy Programme (FIRE), COSP and the Solar Demonstration Programme. For biomass energy, the FIRE programme established in 1979 and scheduled to expire in 1988 was designed to encourage the substitution of mill and forest wastes for purchased energy in the forest industry with a taxable grant of 20% of approved capital costs. In addition to the forest industry, programme eligibility was expanded to include other industries and commercial and institutional establishments. Total programme funding will reach CDN $94 million for 180 projects. This programme is expected to account for an additional 1.36 Mtoe of renewable energy use. The renewable energy sources component of the COSP has allocated a total of CDN $128 million of grants for approximately 220 000 residential conversions to wood-heating appliances. The amount of oil displaced, assuming 50% displacement, would be about 0.45 Mtoe per year. The COSP programme was terminated on 31st March 1985 due to federal expenditure reduction measures. In terms of solar energy, a five-year solar RD&D programme approved in July 1983 provides declining grants for solar installations with the purpose of encouraging the achievement of cost-effective active solar systems by 1988. The programme also has budgeted funds to continue the research and development of solar technology.

In Italy recent legislation provides for grants from 25% to 50% of costs of energy conservation and renewable energy investments in the building, industrial and agricultural sectors. At the local level, the project selection procedure is computerised and approximately 100 000 requests have been made for grants totalling US $270 million. At the national level grants total US $320 million.

In the Netherlands, a new stimulation programme, named Integrated Wind Programme, was initiated in 1986 by the Ministry of Economic Affairs. It comprises two financial instruments aimed at wind turbine manufacturers and at buyers of wind turbines. Wind turbine manufacturers can be supported up to 70% of their development costs with a maximum of NLG 7.5 million per manufacturer. The buyers of wind turbines in the commercial sector can be subsidised with a maximum of NLG 700/kW installed capacity in 1986 declining to NLG 100/kW in 1990. The total budget allocated for the integrated wind energy programme is

NLG 105.5 million for the period 1986-1990. The aim of the integrated wind energy programme is to install 100 to 150 MW wind power capacity in 1991 and to enhance the introduction of commercially-viable wind turbines. To support these developments a national research programme will be executed with a total budget of NLG 21 million for the period 1986-1990. To promote the installation of wind turbines and solar water heating in municipalities and other non-profit institutions (e.g. schools, churches, swimming pools and health services), the Netherlands allocated a direct 30% subsidy, with a minimum of NLG 10 000 investment (expected to become NLG 5 000 within a year).

In Spain, the Energy Conservation Law established that grants could be given for low-temperature solar heating and cooling installations provided that the solar collector is manufactured in Spain and satisfies a specific performance and quality standard and that the other components of the installation comply with certain requirements. Results of this scheme for 1982, 1983 and for 1984 are as follows:

	Budget (million pesetas) [1]	Collectors Installed (m²)
1982	100	15 400
1983	139	21 400
1984	158	22 500

1. On average in 1982, Pesetas 100.00 = US $0.91, in 1983 = US $0.69, in 1984 = US $0.62.

Outside the IEA, in Finland, since 1979 the Ministry of Trade and Industry has encouraged the commercialization of biomass energy applications mainly through a subsidy scheme. Grants are provided to industry and community district heating plants. Subsidies may be granted to small- and medium-sized plants with upper limits of 5 MW in industry and 10 MW in district heating plants. Bigger plants have also been eligible for interest rate support since 1983. The investment subsidy is 20% of the total investment cost of the plant. Since 1985 new technologies could receive up to 40% subsidy, and solar and wind energy systems and heat pumps were made eligible for these grants as well. The subsidies granted were 65 million Finnish Markka (FMK) in 1979 and FMK 127 million in 1981. After that, the amount of money budgeted was decreased. In 1986 there was a total of FMK 40 million available for these subsidies. According to a study made, these grants were of vital importance for the realisation of one half of all the biomass projects in Finland.

Examples of Loan and Loan Guarantee Programmes: In Japan, the commercialization of solar energy is supported by loans of 100 000 yen per house made available by the Housing Loan Corporation for the installation of solar hot water supply equipment in newly built residences. Since fiscal year 1983, additional loans of 500 000 yen per new residential building have been made available by the Housing Loan Corporation for the installation of passive solar systems.

The United States Department of Energy manages a US $500 million Geothermal Loan Guarantee Programme and a US $240 million alcohol fuels loan guarantee programme to encourage and assist the private sector in accelerating the utilization of geothermal and biomass energy. The geothermal programme supports both power generation projects and industrial direct heat applications.

CONCLUSIONS TO CHAPTER IV

Just as governments find it necessary to influence energy supply development from a policy standpoint, they also find it advantageous to conduct activities to promote their development. Since the 1973 oil price increases, government policies and actions have been instrumental in accelerating the development of renewable energy technologies. The main reasons for government involvement were that renewable energies could contribute to energy security and are environmentally benign. Some governments have recognised the additional economical benefit to their national economies from locally-produced renewable energy and from the manufacture of components and equipment within their countries to serve a growing market, both within their borders and as exports.

While government activities can take a variety of forms, ranging from direct funding of R&D of future technologies to actions to facilitate market penetration of economic technologies, the most successful government renewable energy technology development and commercialization programmes were those: 1) that were designed to match the government activity with the stage of development of the technologies as well as with the governments' criteria for meeting its energy supply priorities and 2) that provided long-term commitment and continuity together with appropriate policy and/or funding support with special attention to the timing of government action and evaluation of results.

Government activity in funding R&D for renewable energy technologies has been very significant. The support, over US $6 billion cumulatively by the IEA countries through the end of 1984, and its duration has had the combined effect that renewable energy technologies have been continually advancing technically, increasing their contribution to meeting some countries' energy requirements and contributing to the goals of diversifying their energy supplies.

Governments have now acquired good experience in the renewable energy R&D field which allows them to be selective in their efforts, according to their national circumstances and their resource allocations. For the most part, government RD&D budgets already reflect an appropriate, directed allocation of funds to R&D for technologies matched to the resources indigenous to their countries.

Government incentives and regulations allowed an array of renewable energy technology industries to emerge sooner than they would have without this government involvement. Although this process was often difficult and sometimes inefficient, it did help create the commercial capabilities of the renewable energy technology industries. Development of data bases, standards and codes, providing public information and training, and demonstrating technologies in government facilities are all appropriate and effective means for government action when technologies have reached the stage of early market penetration. As technologies that are now in the "Under-Development" stage become ready to enter the market, these types of activities will accelerate their penetration as well. The same is true for the financial and fiscal incentives that have been provided by governments. Much has been learned about the attractiveness of different incentives to the market as well as about the likely spin-off effects. These lessons should be applied to the design of any new incentive programmes.

Total funding for government programmes has been substantial. Nevertheless, even after a decade of government activity, many renewable energy technologies have still not reached the stage where market penetration can be achieved solely through market forces, and additional activity may still be necessary and justifiable. While, on the one hand, well-designed government policy actions, RD&D funding, and fiscal financial incentives have, for the most part, been found to be effective in promoting development of one or more renewable energy resources and related technologies, governments should not, on the other hand, promote inappropriate technologies. Any national re-evaluation should be based first on technical status today as well as prospects for future technology advances.

No single government action is appropriate to all resources and technologies; the key to maximum effectiveness is timing. Government support should focus on conceptual and applied research, then shift to cost-shared technical development with industry, and finally shift to encouraging commercialization. New concepts for renewable energy technologies are certain to be brought out as work on present concepts matures to development and demonstration stages. The RD&D infrastructure must remain in place, ready to pick up and evaluate these new concepts. It is for member governments to decide in the light of their energy circumstances and policies about the level and type of support to give to the future development of renewable energy.

CHAPTER V

OPPORTUNITIES FOR FURTHER
INTERNATIONAL COLLABORATION

Member governments of the IEA acknowledge a long-term obligation to ensure continuity of energy supplies at acceptable prices. The realisation of this objective involves a diverse programme of activities to develop alternative sources of energy, of which renewable energy sources have been an important part.

While some government R&D for renewable energy technology development to date has involved bilateral and multilateral international co-operative programmes, most has been conducted on a national basis. However, with the recent tighter budgetary constraints on national energy RD&D program- mes, the long-term viability of many national renewable energy technology development programmes could become endangered. Thus, a greater pooling of effort through international co-operation could provide a practical solution to maintaining long-term momentum.

Co-operation in renewable energy technology may have benefits beyond the contributions to national energy balances. For countries with relatively modest technology bases, such co-operation can result in an upgrading of domestic technologies. Also, multi-national co-operation increases contact among researchers who then benefit from experiences in other countries. In addition to the IEA, there are a number of international organisations or agencies working in the field of renewable energy technologies. Examples are the Commission of the European Communities and the International Solar Energy Society. Continued and even greater interaction and co- operation of IEA with these organisations would be mutually beneficial and should be encouraged.

In addition, there are substantial opportunities for enhanced co-operation among IEA Member states via the IEA programme for international RD&D collaboration. A major component of this programme has been the development of renewable energy technologies. For participants in these collaborative activities, the organised exchange of experience and scientific and engineering talent has provided results, data and comparisons, based on a wide variety of options under active development in various countries and has helped to reduce duplication, as well as technical and financial risks. The present range of IEA programmes in the renewable energy technology area covers:

— active and passive solar heating and cooling;

— a unique facility in Almeria, Spain to evaluate different solar thermal power systems;

— the establishment of the world's first man-made geothermal heat source in New Mexico, USA;

— systems studies and practical demonstrations with large-scale wind generators;

— experiments with rapid production and conversion of forestry crops to solid and liquid fuels;

— production of hydrogen from water;

— wave power device development; and

— the organisation and dissemination of technical information on a wide range of biomass technologies.

The mode of co-operation has also taken a variety of forms. One is the pooling of experience through task-sharing. This option introduces the need for rational planning in the allocations of work among participants and at the same time avoids the outflow of funds across national boundaries. A typical example of international task-sharing is linked to the development of large-scale wind energy conversion systems (LSWECS). The participating countries, active in the development and construction of one or more large facilities of approximately 1-3 MW each, are sharing information through co-operation in the planning and execution of their national efforts in the design, construction and operation phases. In addition, a reporting system for operational events has been initiated for quick and accurate information about abnormal events arising from operation, thus facilitating and broadening experience from prototypes.

Joint funding has been adopted in a few cases in achieving the design, construction and operation of costly facilities which might otherwise have been prohibitive for a single country. The installation of experimental facilities near Almeria, Spain through the Small Solar Power Systems project (SSPS) is an example of this type of collaboration. Nine Member countries have funded the installations, constructed on a common site, and jointly carried out the test and evaluation work. Joint funding has also been adopted in the establishment of central information centres concerned with the collection, storage, retrieval, analysis and dissemination of highly specialised technical data, such as the Biomass Conversion Technical Information Service in Dublin, Ireland.

Although international collaboration may offer access to more comprehensive results and wider-ranging experience for less cost, some national programme managers may be hesitant, being reluctant to lose complete control over an activity. Industrial researchers may likewise be reluctant to share information and technical data with potential international competitors. IEA experience shows that it is possible to overcome these reservations by careful planning and definition of tasks and distribution of work including clear provisions in contractual agreements. Also, from the outset there needs to be an understanding about dissemination and application of the research results. Where substantial expenditures from common funds are involved, rules, procedures and authorisation must be established on how to use those funds for the project purposes. With these guidelines IEA international RD&D collaboration has helped to develop experienced management techniques, supported by supervision at a high level by the participating national bodies.

In the present budgetary climate, funds for the design, construction and operation of major new facilities for the demonstration of renewable energy technologies are likely to become increasingly scarce, both for national and international projects. Future international collaboration could thus concentrate on those R&D areas with the most promise for governments interested in additional international collaboration as a means to reduce costs. While it is beyond the scope of this report to develop a list of these priority areas, Table 5, which presents a compilation of R&D needs identified in Annex I, could serve as a starting point for developing additional high priority collaborative projects. In addition, future collaboration could embark on a project of an evaluation of past experiences in collaborative projects and of where there may be needless duplication of effort which could be reduced by international collaboration. Identification of basic problems and assessment of performance, application experience and implementation strategies would also be of great relevance for any new collaborative projects on renewable energy technologies development.

As a next step, the organisation and structuring of workshops, early consultation on the planning and priority of RD&D programmes, and the establishment of a data base for renewable energy sources could be highly productive. The overall objective would continue to be an optimal use of national resources, and international information exchanges could offer some guidance to national governments. Future international collaboration should also find ways and means to allow the private sector to participate more directly in government-sponsored collaborative activities and to play a more important role in the execution of international R&D projects. Such participation will be crucial to the commercialization of technically successful R&D efforts.

CONCLUSIONS TO CHAPTER V

Based upon past experience and recognising the need for greater selectivity in future renewable energy technology development, international col-

TABLE 5
RD&D ACTIVITIES NEEDED FOR RENEWABLE ENERGY TECHNOLOGIES

Active Solar
Increased collector efficiency (performance, reliability, durability, lifetime)
Improved materials (cost and mass reduction, reliability)
Long-term storage
Solar cooling collectors
Systems design and intelligent systems control

Passive Solar
High performance building components and materials (variable transmittance and insulating glazings, phase change materials with high thermal storage capacity as building materials)
Convection systems
Cooling systems
HVAC and lighting system controls
Daylighting systems (new materials, light wells)
Computer-aided design tools (simulation programmes for large-scale detailed calculation and simple microcomputer application)
Integral systems development

Solar Thermal
Materials and components (improved performance and reliability, mass and cost reduction by stable lightweight structures; absorbtive surfaces, radiation conversion, new heat transfer media (oils), receiver concepts, heat engines, concentrators)
Systems and component analysis, lifetime prediction and design
Systems control
Heliostat components cost reduction (silver/steel substrates, stressed membranes)
Direct absorption receivers
Flux concentration in control receivers
New thermo/photochemical processes for storage or for transportable high-density fuels
Demonstration of components and full-scale systems

Photovoltaics (PV)
New PV materials (basic research for improved efficiencies and cheaper materials)
Reduced production and system costs (new processing techniques/methods, concentrating systems, batteries and power conditioners)
Cell, module and systems design (improved efficiency and mass and cost reduction by thin film/multi-junction design)
Site-specific resource assessment
Demonstration (pilot) plants

Wind
Small and large systems components (materials, blade design, airfoils, controls drive trains)
Variable speed generators (pitch control, wall regulation, feathering strategies)
Aerodynamics/atmospheric physics (turbulences, materials fatigue, rotor weight)
Resource assessment (velocity measuring techniques, wind atlases, modelling)
Grid connection and load following
Wind-diesel hybrid systems
Offshore wind stations
Heat production
Noise reduction
Application of theoretical knowledge to experimental wind turbines

Biomass
Biomass growth, harvesting and production techniques
Genetic basic research and engineering
New sugar/oil bearing plants
Advanced cellulose and lignocellulose conversion techniques (cost reduction of ethanol production)
Special fields in gasification, liquefaction and pyrolysis
Aquatic biomass
Fermentation and steam explosion techniques
Biochemical and microbiological reactions
Ecological and environmental aspects of energy farming and biomass combustion

Geothermal
Resource prospection and exploration techniques
Resource assessment and reservoir behaviour modelling
Drilling systems and components (hard rock, directional and magma drilling)
Materials (corrosion and scale prevention)
Corrosive brine handling
Brine disposal injection techniques
Power conversion systems (large binary cycle, high saline conditions)
Hot Dry Rock (fracturing techniques, frac-localisation,)
Geopressured reservoirs
Magma technology

Ocean
Demonstration/testing of full-scale wave energy systems
OTEC heat exchangers with biofouling control and cleaning techniques
OTEC cold water pipes materials
Demonstration full-scale open-cycle OTEC plants

Energy Storage
Thermal storage techniques
Electrical storage techniques

Electricity Transmission System Interfaces
Development of equipment to handle power supplied by intermittent renewable energy sources

laboration would probably be the most fruitful if it were to concentrate its efforts on the R&D areas with the most promise for the participating governments. Prioritising new areas of R&D interest is a first step in the process of concentrating these efforts. In addition, a number of other areas of international activity which have been identified during the development of the information for this report could be considered for activities involving IEA and other international organisations with similar interests. These are:

— Creation of a forum for industries and governments to share the experience in technology transfer, information dissemination and commercialization, such as an IEA Workshop on commercialization.

- Development among economists of common methodology and assumptions for analysing the economical potential of renewable energy technologies, such as an IEA Workshop on economic analysis techniques for renewable energy technologies.

- Evaluation of which government policy actions and incentives have been the most effective in accelerating the commercialization of the individual renewable energy technologies.

- Development of a common, acceptable methodology for estimating the long-term contribution of renewable energy sources.

CHAPTER VI

CONCLUSIONS AND RECOMMENDATIONS

Renewable energy sources presently provide significant amounts of energy in a few IEA countries. These energy sources hold promise for a substantial potential contribution for many IEA countries, and they will therefore become increasingly important in the future. However, this study finds that at present it is still quite difficult to estimate the exact size or timing of this contribution. This is because, with the exceptions of hydropower and some biomass technologies, many of the related technologies are still under development or have only recently been commercialized. In the field of energy, twelve years is a very short development period, and much more experience is needed to predict the future economics and markets for these emerging technologies. Many of these renewable energy technologies will require substantial additional development or market penetration before they can make significant contributions to national energy supplies. For the present, and for some renewable energy technologies, possibly for several decades to come, renewable energy technologies will probably enter the market or increase their market share primarily in areas where the costs of conventional fuels are highest or in areas where resource availability and site conditions are particularly attractive. This is not only because the related technologies are relatively immature but also because they are capital intensive. Moreover, prices for conventional energy sources have increased less than previously expected, and the short-term attractiveness of some renewable energy technologies has diminished somewhat. This means that only the most optimum renewable energy installations can presently compete in economic terms with conventional energy technologies.

Up to twelve years of development efforts by governments and industry has resulted in substantial progress in the case of each renewable energy technology studied and substantial information on the capabilities and limitations of each technology. Because of the major differences in the types of renewable energy sources, the stage of development of their related technologies, and the factors which affect the continued progress, these energy sources cannot be considered and discussed as one energy source. Rather, categorization of their technologies according to their stage of technical development helps to clarify how and where certain factors become most critical. Dividing the technologies into four stages, "Economic", "Commercial-With-Incentives", "Under-Development" and "Future-Technology", provides a framework for discussion.

While all technologies can always benefit from further technical refinement, technologies in the "Economic" and "Commercial-With-Incentives" stages would generally benefit most from mass production and/or economies of scale. Mass production could substantially lower the cost of some technologies and thereby positively influence their market penetration. The renewable energy technologies most dependent on mass production for cost reduction, at present, are active solar space and water heating systems, small- to medium-size wind systems, parabolic troughs and small photovoltaic systems. Economies of scale and capital availability for very large projects become critical for other technologies. Tidal power is the most notable technology whose very large economic size may require innovative financing arrangements in order to expand the number of commercial projects. In addition, costly and lengthy site-specific resource assessments are critical to obtain financing for small to medium wind systems and geothermal electricity generation and direct heat. There are several other institutional or market factors which appear to critically affect "Economic" or "Commercial-With-Incentives". Geothermal direct heat is probably the most geographically-limited technology which presents locational challenges for full utilization. Technologies which produce heat directly to an end-user, such as a residence, are subject to information dissemination and diffusion problems similar to those encountered by conservation technologies. The renewable energy technologies most susceptible to this problem and whose development is primarily limited by it are: passive solar heating, daylighting, biomass (heat and anaerobic digestion) and geothermal direct heat (in addition to geographic problems).

Negative environmental impacts can be a constraint on development of an "Economic" technology, and, while most of the renewable energy technologies are relatively benign environmentally (i.e. environmental impacts are minor and controllable), biomass combustion technologies can produce environmental impacts to air and water. In addition, efforts to increase biomass availability (collection techniques, energy cropping), raise concerns in land-use, agriculture, and forestry policy areas.

Many others of these technologies are in early developmental stages and will require extensive RD&D before they can be economically competitive with conventional energy systems, especially with present electricity grid networks. For the renewable energy technologies which are not in the "Economic" category, RD&D successes will be the critical factor in advancing the technology toward commercialization.

The development of renewable energy resources in general will require support from government as well as industry especially for the technologies which have not achieved significant market penetration. Industry cannot be expected to carry the full burden of pursuing costly development projects for which the economic and commercialization outlook is uncertain, technical risks are high and pay-back periods are long. For the many renewable energy technologies which have considerable potential to contribute to energy needs but which still need additional technical development, it will be important for governments to continue R&D efforts. New renewable energy technology concepts will undoubtedly be developed in the future. Without the infrastructure in place to develop them to the point where their feasibility is determined, additional potential from renewable energies will be much more slowly realised.

Governments may also need to play a facilitating role in establishing the necessary institutional and commercial environments to accelerate the introduction of those renewable energy technologies which show promise of proving economically viable at an early date. At the commercialization stage, policy measures should be carefully planned within the context of general energy policy. There is a need to remove barriers to increased market penetration for technologies which are economically viable. In pursuing these roles, policy makers should consider that the development of viable renewable energy options will be a long-term endeavour, requiring balanced and consistent support. They will also need to take care not to introduce an unintentional bias in favour of renewable energies which would prejudice the further economic development of other alternatives to oil or the exploitation of the indigenous energy resources of IEA countries or vice versa. Moreover, all countries may find it necessary to be more selective to get the maximum results from the limited human and financial resources, based on experience and technological advances made during the last decade. Areas of concentration will vary from country to country, depending on different energy situations, resource allocations, and research and development priorities. Therefore, it will be necessary to establish priorities and make choices at regular intervals in order to give the proper emphasis to the most promising technologies based on their technical merit, potential contribution, economic competitiveness and environmental impact. Periodic programme evaluations can provide a framework for achieving greater selectivity and effectiveness of government programmes.

RD&D work by governments on several renewable energy technologies has been critical to making them economic. IEA co-operative RD&D efforts were instrumental in this respect and should be strengthened as well as continued.

International collaboration can help to reduce the cost burdens of individual governments and industries and thus encourage increased participation in renewable energy projects. Further international co-operation should have several aims:

— For technologies that are economic or nearly so, the international exchange of experience and results could stimulate more developments.

—   For technologies under development and for future technologies, it will be less costly for individual countries if this work is carried out and shared internationally and, therefore, it will be more likely that developments are continued long enough to get them to market readiness in the first case or to determine their feasibility for the second case.

Both industry and government share the objective of reducing energy costs and increasing security of supply over time. Close co-operation between government and industry is essential. While industry and government may not have similar ideas about which areas are worth pursuing, mutually-favoured areas for continued research, development and demonstration of renewable energy resources are most likely to be those which have a definite potential for cost reductions.

The specific recommendations of this review are that governments should, as appropriate to national circumstances:

—   maintain long-term RD&D efforts supported by evaluation of past government programmes and activities and of economic and technical progress of the technologies targeted since continuity is indispensable for introdution of new renewable energy technologies into the market, and seek broader participation of the private sector in government-sponsored collaborative activities;

—   encourage the development of an economically-viable renewable-energy-technology industry and the development of innovative financing arrangements for economic investment in renewable energies;

—   seek, as a matter of general energy policy, in accordance with agreed IEA objectives, to avoid and remove distortions in the energy market, particularly in fiscal treatment of energy supply investments;

—   ensure, either directly or through their influence with the appropriate regulatory authorities, that electric utilities base their policies for buying and selling electricity to renewable energy producers on avoided costs;

—   carefully plan and design policy actions supporting the development of renewable energy technologies, such as the adoption of standards and codes or the modification of utility regulations, the development of data bases and reliable and appropriate information for potential users of renewable energy devices, financial incentives and the removal of institutional barriers which inadvertently block renewable energy technology commercialization;

—   provide timely information and education programmes for both users and installers of renewable energy technology, also sustaining new creative financial and diagnostic services for promoting these investments; and

—   concentrate international collaboration on:

- development of specific components where the technology still has bottlenecks to be broken through;

- workshops on the problems of commercialization and on economic analysis of renewable energy technologies;

- stimulation of renewable energy resource "prospecting" by private industry, and work on developing a common methodology to estimate future contributions from renewable energy sources; and

- technology transfer.

NATURE, STATUS AND OUTLOOK OF RENEWABLE ENERGY SOURCES AND CONVERSION TECHNOLOGIES*

CHAPTER ONE: INTRODUCTION

The first annex to the review of *Renewable Sources of Energy* attempts to function as a "technical yardstick" by which to assess renewable energy sources and technologies. To do this, the annex provides detailed descriptions of the nature of renewable energy sources (solar, wind, biomass, geothermal and ocean energies), includes their potential and availability in IEA countries, the physical and thermodynamic principles which allow the resources to be utilized, and general rules-of-thumb on how the resource potentials are measured.

This annex also presents detailed descriptions of the technologies which have been and are being developed to "tap" each renewable energy source. Each technology·description covers:

- a description of the equipment and means by which the resource is converted to usable forms of energy;

- the various applications which have been developed or identified for the energy produced;

- the technical and commercialization status of the technology and the major recent accomplishments to improve either its efficiency or to prove the concept or feasibility of the technology, depending on its status;

- environmental considerations resulting from the use of the technology;

---

* Preparation of this Annex was substantially assisted by Ms. Sheila Blum of International Planning Associates, Inc., Beltsville, Maryland, U.S.A.

- the outlook; and

- major findings which summarize the status and prospects for the technology. (These findings are also used in Chapter II of the main body of the report).

The costs of a technology are a very important consideration in determining its status and outlook. Examples of costs are provided in the appropriate sections for illustration of either past accomplishments in reducing costs, or predictions of how costs might be reduced to change the commercial outlook of a technology. It should be noted that Annex II to this report comprises the economic analysis which is used to provide an "economic yardstick" by which to measure a technology's comparative competitiveness with conventional energy. Thus, the costs used in Annex II have been "normalized" for comparison purposes; whereas cost figures used in this annex *may not* be used for comparison purposes.

Several points should be made here because they are common to all of the renewable energy technologies discussed. First, in all cases, renewable energy sources had been used at least to some degree in the past. Most of the technologies were understandably much more primitive or unsophisticated than those that have been under development recently. The decline in use primarily stemmed from oil and natural gas supplanting renewable energy technologies around the turn of the century. Cost advantages and the ease of use of oil and gas were the motivation for their replacement. Renewed interest in using renewable energy sources and technologies occurred in the early 1970s caused by oil shortages and price shocks. Therefore, most of the discussion herein on accomplishments focuses on the period of renewed development from 1974 to the present.

CHAPTER TWO: SOLAR ENERGY

A. NATURE OF THE SOLAR ENERGY SOURCE

The sun is a constant and primary source of heat and light on Earth. At the outer atmosphere the solar energy constant is 1 373 W/m². This energy is transmitted as radiation waves ranging from low-energy long radio frequencies to higher energy X-and gamma rays. Radiation in the visible light band makes up 46% of the sun's radiation, while 49% is in the infrared band, sensed as heat. The remainder is in the ultra-violet band.

The atmosphere affects how much radiation and what type of radiation actually reaches the surface of the earth. During periods of heavy clouds or fog, essentially all the radiation reaching the ground is diffuse, that is, completely scattered by particles and molecules in the air. On a clear day, much of the radiation that reaches the surface is unscattered and is called direct or beam. On the surface, the total or global radiation received by any object is the sum of the direct and diffuse components, and is called insolation (or hemispherical insolation). The maximum amount that is available at the earth's surface on a clear day is normally about 1000 W/m².

The difference between the amount of energy at the outer atmosphere and the planet surface, approximately 25%, is due to scattering by air molecules and dust and absorption by water, carbon dioxide and other airborne matter. The amount of insolation available for conversion to energy at any location depends on several key factors: location of the sun in the sky (altitude and azimuth), which varies daily and seasonally; atmospheric conditions, both general and in the micro-climate; altitude above sea level; and hours of daylight. Besides weather patterns, radiation resources at a site may also differ due to air pollution, either from man-made or natural sources such as volcanoes and intense dust storms. Although total ground-level radiation may vary by as much as 25% or more annually, it is fairly constant in any one location.

The process of determining the potential contribution of solar resources involves determining the amount and type of radiation available at a site. Atlases are available that provide the following information for each month:

Average daily global solar radiation on a horizontal surface
Average daily global solar radiation on tilted surfaces
Average daily direct normal radiation
Average daily direct normal radiation on tilted surfaces
Average daily diffuse solar radiation on a horizontal surface

In addition, atlases may provide weather data of use to system designers:

Average daily opaque sky cover
Normal daily average, minimum and maximum temperatures
Normal total heating and cooling degree days
Average daytime and night-time dry bulb temperatures
Average daily wet bulb temperature

The data available in an atlas are based on measured data at sites spread throughout geographic regions. The ideal situation is to have long-term data measurements available at the location being assessed.

FIGURE 3

ANNUAL GLOBAL SOLAR RADIATION ON TILTED SURFACE

(Range)

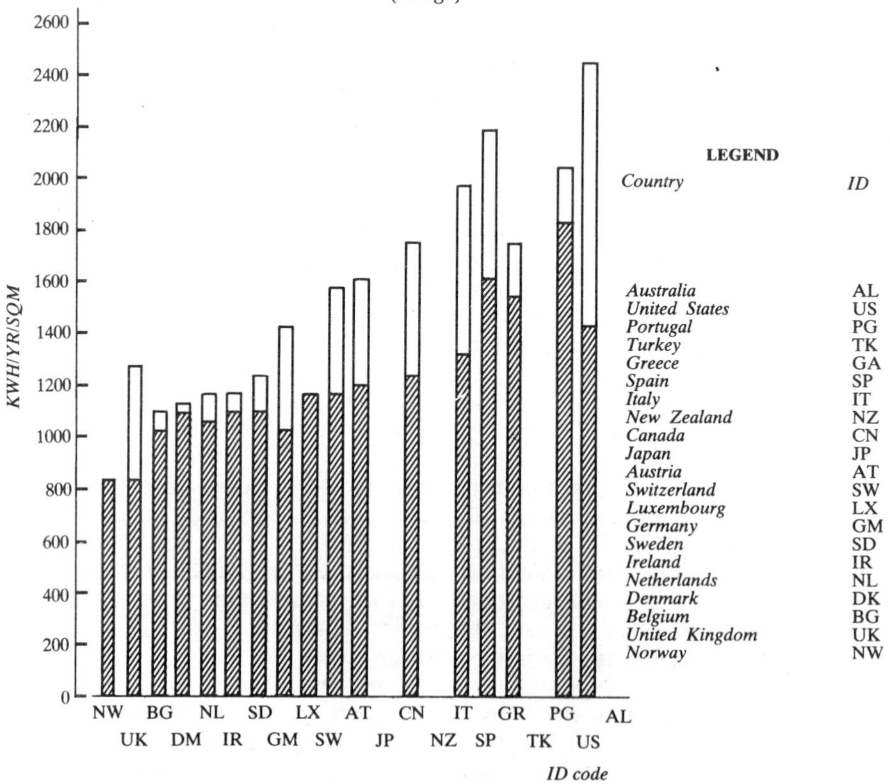
2600
2400
2200
2000
1800
1600
1400
1200
1000
800
600
400
200
0
KWH/YR/SQM
NW
UK
BG
DM
NL
IR
SD
GM
LX
SW
AT
JP
CN
NZ
IT
SP
GR
TK
PG
US
AL
ID code
LEGEND
Country
ID
Australia AL
United States US
Portugal PG
Turkey TK
Greece GA
Spain SP
Italy IT
New Zealand NZ
Canada CN
Japan JP
Austria AT
Switzerland SW
Luxembourg LX
Germany GM
Sweden SD
Ireland IR
Netherlands NL
Denmark DK
Belgium BG
United Kingdom UK
Norway NW

Source: United States Department of Energy.

Solar Energy Resources in the IEA Region

Figures 3, 4 and 5 present key parameters that describe the nature of the solar energy resource in the IEA member countries. The first figure indicates the average global or total solar energy incident on a collector that faces south (or north in the southern hemisphere) and is tilted up from the horizontal at an angle equal to the latitude of the site. This collector orientation often maximizes annual collection of radiation for fixed collectors, and is typical of active and photovoltaic systems. The figure illustrates the highest value for each of the countries. The lowest values are found in northern and coastal countries and the highest values in Australia, southern Europe and the United States.

Figure 4 shows the portion of solar radiation that is direct beam. This is the only insolation component that concentrating collectors for solar thermal systems can use and the most important one for photovoltaic systems. These

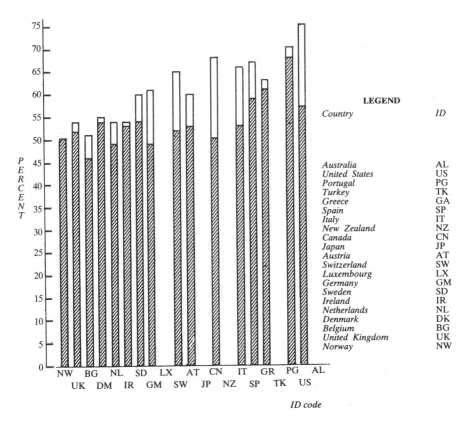

FIGURE 4
ANNUAL DIRECT (PERCENTAGE OF GLOBAL) SOLAR RADIATION ON TILTED SURFACE
(Range)

LEGEND

Country	ID
Australia	AL
United States	US
Portugal	PG
Turkey	TK
Greece	GA
Spain	SP
Italy	IT
New Zealand	NZ
Canada	CN
Japan	JP
Austria	AT
Switzerland	SW
Luxembourg	LX
Germany	GM
Sweden	SD
Ireland	IR
Netherlands	NL
Denmark	DK
Belgium	BG
United Kingdom	UK
Norway	NW

Source: United States Department of Energy.

technologies require relatively high values of direct beam radiation, which is most available in sections of Australia, the United States, southern Europe and Switzerland.

Figure 5 presents a measure of sky clearness. A low value, e.g. .30, indicates a country with cloudy sky conditions, while a high value, e.g. .60, indicates a high level of solar potential. These figures reveal the disparity in resources that exist from one country to another and explain why certain solar technologies are emphasized more in R&D programmes or have begun to capture market shares in some countries, while other renewable energy technologies (bio-energy, wind) will be more emphasized in others.

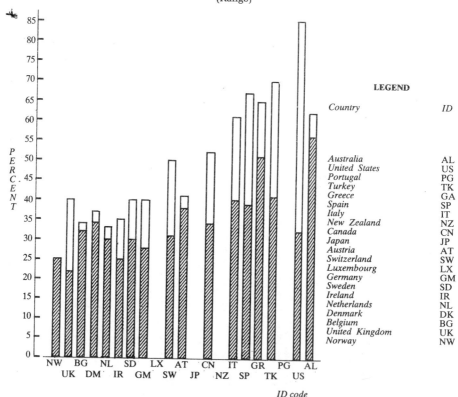

FIGURE 5
ANNUAL SUNSHINE PERCENTAGE OF POSSIBLE
(Range)

Source: United States Department of Energy.

It should be noted that for photovoltaic (PV) systems, the conversion process is more complicated and exact knowledge of the distribution of the energy content in sunlight is required as solar cells respond differently to different wavelengths of radiation. The semi-conductor materials that are used to make solar cells are fabricated so that they have band-gaps in their electron structure that allow an incoming radiation photon to displace an electron in the material from its hole-pair and cause electrical current. A portion of the incident radiation will not have sufficient energy to activate the cell and another portion will have too much energy and be converted to heat or reflected away. PV cells can convert both diffuse and direct radiation.

B. ACTIVE SOLAR HEATING AND COOLING TECHNOLOGIES

I. INTRODUCTION

The antecedents for active solar energy systems can be traced to the 17th century in northern Europe when greenhouses were introduced to protect tropical plants brought home by explorers. In this period, a French scientist, Horace de Saussure, began experimenting with black boxes covered with glass to elevate the temperature of collected solar radiation, thus inventing the "hot box". It was not until the late 19th century in the United States, however, that the principle of collecting solar energy was developed into a commercial product: solar water heaters.

The first commercial solar water heaters were simply bare water tanks placed in the sun. These soon gave way to a more efficient approach, enclosing the tanks in a box with a glass cover. Then in 1909, an American named William J. Bailey separated the collector from the tank storing the heated water, an innovation that would be followed by more than seventy years of development and refinement.

These earliest water heaters are examples of "passive" technology, i.e. they did not incorporate any mechanical energy to move fluids or deliver the captured heat. This idea may have been first developed at the Massachusetts Institute of Technology in 1939, when Dr. Hoyt Hottel installed pumps to circulate fluid through flat-plate solar collectors for an experimental home heating system. Five years later, Dr. George Lof built a system in Colorado, United States, using electric fans to circulate air through solar collectors, also for home heating. These and other systems began to establish a technical approach of using stationary flat-plate collectors with an active means of removing and storing heat for later use which has become the basis of the modern active solar heating system.

II. TECHNOLOGY DESCRIPTION

When metal or other materials are exposed to photons in solar radiation, the molecular structure absorbs a portion of the energy, increasing the temperature of the material. It is this molecular process that active solar systems use for conversion of solar radiation to useful energy. The absorbers that are used are able to convert both beam and diffuse radiation to heat.

The amount of solar energy converted depends first on how much is intercepted by a collecting device, then the physical characteristics of the collecting device and its interaction with the surrounding environment, the load and any storage sub-system incorporated in the system. Active systems normally do not employ tracking devices because flat-plate collectors can convert both diffuse and direct radiation and can deliver energy at temperatures high enough for most heating and cooling applications. The collectors are usually mounted in a stationary position and set at a certain tilt angle and east-west orientation in order to collect the most energy for the application. This orientation is usually determined by seasonal needs.

The term "active" solar system differentiates this technology from the earlier solar water heaters that employed no mechanical means to move the heat from the collector to the point of use or storage sub-system. These older "passive" technologies (batch and thermosiphon water heaters) still exist, and the companies that manufacture them are frequently included when the active solar industry is described. "Active mechanical means" to move energy refers to use of a pump in hydronic (liquid) based systems or blowers in air based systems.

A typical active solar heating and cooling system consists of one or more collectors, an energy transport system to move that heat to the point of use, an electronic control system, and an energy storage system. Besides operating in temperature ranges that satisfy building energy needs, active solar technology can supply heat for industrial processess at moderate temperatures.

Solar Collectors and System Components

The essential component in an active solar system is the collector. Three basic types have been extensively developed: flat-plate, evacuated-tube, and non-imaging and imaging concentrator. The most common type is the flat-plate, and it is the one most used in active heating and cooling systems. The other two types are commercially available, but have not yet been mass produced.

Flat-Plate Collectors. Flat-plate collectors accept total insolation, direct and diffuse components, and convert it to heat through the thermal reaction of the absorber to the incoming radiation. Thermodynamic principles are involved that predict thermal performance based on the conduction, convection and radiation properties of the absorber. When glass or plastic covers are used, their ability to transmit solar radiation is an important factor because glass blocks the passage of some radiation, and its interaction with the absorber determines the actual conversion efficiency of the collector. The modern flat-plate collector consists of seven major parts:

1. A cover plate or glazing
2. An absorber plate
3. A coating on the absorber plate
4. Passages for the heat-transfer medium
5. Insulation
6. A housing or box
7. Gaskets and seals

Figure 6 shows a cross-sectional view of a flat-plate collector

FIGURE 6
CROSS SECTIONAL VIEW OF A FLAT-PLATE COLLECTOR

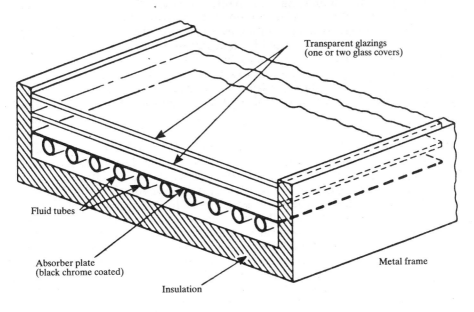

The transparent cover plate acts as a heat trap, allowing solar radiation to pass through to the absorber but blocking the transmission of re-radiated heat and preventing atmospheric air currents from cooling the collector absorber. Glass and plastic are used for the glazing material, and some designs employ two or three covers to increase the thermal performance of the collector. A trade-off always exists, however, between blocking radiation passage into the collector and blocking the loss of collected heat.

The energy conversion performance of the absorber material can be enhanced by selective coating which reduces the radiative losses. The absorber, usually black in colour, incorporates passages through which the heat transfer media flow to carry heat to the end-use or to storage. The sides and back of the collector are filled with insulation to reduce heat losses. The box is made weathertight through seals and gaskets: however, it must allow condensation to escape and expand and contract through daily thermal cycles without damaging the internal components.

Flat-plate collectors owe their success to their simplicity of construction and their fixed mounting. Their development answered the need for solar water heaters that could survive freezing climates and continue to function. The thermal efficiency of a flat-plate collector depends on its ability to reduce conduction losses in the absorber, thermal radiation losses from the heated parts of the collector to the surrounding environment, and convection losses due to air circulation. As the average fluid temperature in a collector increases, thermal losses increase, and the efficiency declines.

A variation of the flat-plate collector that has been a commercial success is the unglazed, or swimming-pool panel. A key determinent of collector efficiency is the difference in temperature between the collector and the environment. In summer or in climates where no cold season exists, simple unglazed absorbers can provide temperatures well above 25°C with instantaneous efficiencies approaching 80%. The collector consists of a simple metal or plastic panel and has a high flow rate with many channels for the fluid flow. In cold conditions, radiative and convective heat losses severely lower the performance of such panels and a glazed model becomes more efficient.

Evacuated-Tube Collectors. Evacuated-tube collectors, which are made with long glass tubes (1-2 m), incorporate the same basic components as flat-plate collectors and function in the same way, with one major exception. Rather than employ insulation (which is limited to the back and sides of a flat-plate collector) to reduce heat losses, this type of collector uses a vacuum to insulate the absorber surface from convection and conduction losses. The vacuum also protects the absorber surface from degradation. An evacuated-tube collector consists of a set of such tubes connected to a support structure and manifold to carry the energy to the point of use. Several different designs exist for the absorber and the manner in which energy is taken from the tube. The two primary approaches have been: 1) pumping a working fluid through tubes attached to or part of the absorber; and 2) using a heat pipe attached to the absorber to carry heat to a manifold heat exchanger. Some manufacturers have developed tubes made only of glass; others have used metal components for the absorption and heat transfer.

From a thermodynamic standpoint, evacuated-tube collectors compare favourably with flat-plate units. The lower heat loss characteristics allow evacuated-tube collectors to operate more efficiently both under low irradiation conditions and at operating temperatures above 80°C. Like flat-plate collectors, evacuated-tube units are usually mounted in a fixed position. Their capability to operate at significantly higher temperatures, however, allows them to use low-concentration reflectors to increase the collection area and the temperature of collected energy. The tubes or reflectors may also be adjusted seasonally to increase annual energy output.

Concentrating Collectors. Concentrating collectors which are essentially stationary, but which may be adjusted periodically, have been developed using a compound parabolic-concentrator shape proposed by researchers at the University of Chicago. The shape of the reflective surface directs the incoming radiation within a wide angle of acceptance to a small receiver area. This concept has been incorporated into prototype evacuated-tube collector designs.

Tracking concentrators are sometimes used for active solar systems for residential and commercial installations. Smaller versions of single-axis parabolic troughs are the type usually employed. They may be roof or ground mounted and require a sun-position sensor and tracking motor. These systems will only collect energy effectively from direct or beam radiation, limiting their applicability to sunny climates. This is because a reflective or a refractive surface is used to direct the incoming radiation on to a smaller surface. The smaller surface acts as the thermodynamic absorber and operates with the same characteristics as a flat-plate absorber, except that the temperature regime is much higher. Concentrators cannot convert the diffuse radiation.

Energy-Transport Systems. Two basic types of energy distribution systems exist: open-loop, and closed-loop. In an open-loop system, the heat transfer fluid is water, and heat is directly delivered as service hot water. These systems are protected from freezing by using a drain-down strategy, in which a temperature sensor activates an electrically operated valve to dump the fluid from the system. Closed-loop systems prevent freezing in three ways. The most common is use of an anti-freeze mixture or non-aqueous fluid in the collector loop with a heat exchanger to transfer the energy to the potable water. Two other approaches are: 1) drainback systems where the thermal fluid, normally water, drains by gravity from the collectors into another part of the system when pumping stops; and, 2) recirculation where the pump is operated during freezing conditions to heat the collectors. Recirculation systems are usually used in areas where freezing conditions only occur a few nights per year.

Air is often the working fluid in residential space heating systems. There is also increasing interest in use of condensible organic fluids. These systems are sometimes called heat-pipe collectors or phase-change units. Organic and silicone oils are also used in some systems as the heat collection and transfer medium.

Heat exchangers are an integral part of most active solar heating and cooling heat-transport systems. They transfer energy from the collection media to storage, and from storage to the load. Many approaches and varieties of heat exchanger equipment are employed. The tube and shell type have been used extensively for hydronic systems.

Solar-System Controls. Active solar systems require sensors and controls to operate their mechanical components. The type of control usually employed is a differential thermostat. Typically, thermistors are placed on a collector plate and in the storage unit to turn the system on and off when energy is collectable. A controller must handle transient conditions, avoid short cycling of pumps, protect against overheating, and take appropriate measures to safeguard against freezing in non-passive approaches (e.g. drain-back). In the very simplest systems a single temperature reading will operate the system in an on-off strategy. In larger systems, centralized control systems often employ micro-processors for more sophisticated operation and control. In addition to controlling the solar system, controls must also integrate the solar unit with the conventional heating system and may also operate the total space conditioning system.

Energy Storage. Active solar heating and cooling systems utilize energy storage to satisfy energy demand at night or during prolonged cloudy periods unless conventional back-up systems are used. The storage system absorbs heat from the collector heat transfer media, stores it until needed, then transfers it to satisfy the load. The most common approach is the use of water containers which may be conventional water heater tanks or other suitable containers made of steel, concrete, fibreglass or plastic. Air-type systems often incorporate a rock bed which typically is a large container with uniformly sized (5 to 8 cm diameter) stone. Insulation of any energy storage sub-system is essential.

In extreme latitudes, seasonal energy storage may be used to furnish winter heating from energy collected during summer months. Generally, these systems are large scale, ranging in capacity from five thousand to one million cubic meters.

III. APPLICATIONS

In the residential and commercial building sector, active solar heating and cooling technology may be segmented into four basic applications: swimming-pool heating, water heating, space heating, and space cooling. In the industrial sector, in addition to these four applications, active solar technology may also be used to provide heat for agricultural and industrial processes.

Swimming-Pool Heating

The simplest application for active solar technology is for heating water to the low temperatures needed for residential or public swimming pools. In the system the pool itself acts as the energy storage system. In most cases, the pool water is passed directly through the panels. Heating systems for outdoor pools are usually operated only in the summer when the ambient temperature is close to the pool temperature. Panels made of bare metal sheet with tubes or inexpensive moulded or extruded polymers or elastomers are typical.

A typical residential swimming-pool system may employ from 15 to 40 m^2 of solar collectors. Public pool systems may employ from 30 to 300 m^2, and, in some cases, this application will use glazed, medium-temperature collectors.

Water Heating

Active solar systems are usually used for heating water in cold climates where freezing is a potential problem. In warmer climates, passive batch type or thermosiphon units are more common. The application where most systems have been installed is in single family residences. For example, fully 80% of residential systems sold in the United States are retrofitted to existing buildings.

In most IEA countries, a residential solar domestic hot water (DHW) system for a single family requires 2 to 3 m^2 of solar collector. Such a system should satisfy from 50% to 80% of annual energy needs.

Active solar water heaters may also be used for multifamily buildings. These systems are usually designed with a large array of collectors and a large storage tank. Commercial and industrial systems are still larger, and designed to meet specific loads. The system design is dictated by load requirements.

Space Heating

Typical residential solar space heating systems in most temperate regions are designed to meet only part of the annual building heating load and a portion of the domestic hot-water requirement. Air-type units have been employed in forced air space-heating systems, but if conventional heating systems are hydronically based, liquid systems are likely to be chosen. Such systems almost always include storage, but usually with only enough capacity to meet demand for a night or one or two cloudy days. In the summer, the system rejects any solar radiation not needed for space or water heating. This lost potential contributes to lower cost-effectiveness of space heating systems per unit of collector area compared to a smaller solar water heater used 12 months a year unless only a comparatively small fraction of the annual heating load is intended to be met by solar. Economics of solar may then compensate for wasted summer heat. The cost effectiveness of combined space heating and water heating systems may be greater than water heating alone because of economies of scale if the fraction of total annual space heating load carried by solar is not too high, commonly below 50%. In some countries at more extreme latitudes, space heating systems with only diurnal energy storage capacity are practically useless since there may be an insignificant amount of sunshine during two or three months of the heating season. In such regions, it is necessary to store energy collected in the summer-time for use during the heating season. Practical schemes have been developed for large-scale seasonal thermal energy storage, and central solar heating plants employing seasonal storage are now economically viable in countries such as Sweden. Seasonal storage may find application in more temperate climates as well because it increases the utilization of solar collectors by a factor of two or three and allows systems with much higher solar fractions.

At the other end of the storage spectrum, direct heating systems with no thermal energy storage have found favour in some markets because their initial cost is low.

Solar Cooling

Active solar technology can be used to operate solar cooling systems. Closed-cycle systems use a conventional refrigeration cycle to extract and discharge heat from a medium. The solar energy system provides the higher temperature heat necessary to drive the cycle. (Consequently, active solar technology that can deliver temperatures well above 100°C is preferred since they increase efficiency. Evacuated-tube collectors are particularly suitable for this application.) High temperature (above 100°C) heat supply is therefore sometimes an advantage but other requirements of the cycle (as in absorption systems) may dictate considerably lower temperatures.

Active solar technology is suitable for both residential and commercial cooling applications. For residential cooling, manufacturers have developed special absorption chillers designed for moderate temperature (below 100°C) input from flat-plate collectors. Commercial solar cooling systems may use absorption chillers designed for gas-fired operation that have been modified to operate with evacuated-tube or parabolic-trough collectors.

Another approach to solar cooling that escapes the limits of Carnot cycle efficiency is to use desiccants with open-cycle evaporative cooling. Desiccant cooling systems pass air through a moisture-absorbing material to dehumidify it, and then cool the air in an evaporation cycle. Solar energy is used to dry or regenerate the desiccant, and the collectors may not be required to furnish temperatures higher than 60°C.

A third approach is to use an active solar system to produce electricity with a low pressure Rankine cycle engine/generator and to then operate conventional cooling equipment. In any case, application of active solar technology for cooling systems continues to have appeal because the cooling load coincides with availability of the solar resource. Additionally, in most climates where space heating is also required, the solar equipment achieves greater utilization and thereby becomes more competitive with conventional energy sources.

IV. STATUS AND ACCOMPLISHMENTS

Status in 1974

In 1974 nothing like the active solar heating and cooling technology industry that now exists was present. In a few IEA member countries, several dozen companies existed that were selling solar water heaters. Many of these were in Japan and Australia and were founded in the 1950s and 1960s. The solar industry in the United States had almost completely disappeared.

In the early 1970s, solar water heating systems were predominantly thermosiphon types. The flat-plate collector produced was usually made of metal pipe fixed to a metal sheet in a carbon steel enclosure with a minimum of insulation and a regular glass cover. The optical and thermal performance of these early collectors was minimal (optical near 65% and thermal conversion near 30% at output temperatures of 65°C). Solar water heating systems for other than domestic applications were very few in number and were usually custom manufactured. A few experimental space heating systems consisted of flat-plate collectors using liquid or air with a water tank or rock-bed storage system. Their thermal energy transport systems used conventional building products, which were frequently ill-adapted for solar applications. Solar cooling existed only as academic experiments. The performance of any of these early systems was rarely measured, and the reliability generally inadequate.

Technical Advances

As a result of national R&D programmes, research conducted by industry, and international cooperative initiatives, active solar heating and cooling technology has advanced in a large number of technical areas. In most IEA countries, greatly improved flat-plate collectors for medium temperature applications now exist commercially. These durable metal/glass configurations have benefitted from advances in absorber design, development of durable, higher performance selective coatings, use of low iron-content glass, improved insulation and weatherization techniques. As a result of these improvements, efficiency of commercially produced flat-plate collectors has improved tremendously. For example, for all collectors submitted for testing to a solar energy test and measured in standardized tests, efficiency improved about 30% since 1977 (see Figure 7).

Industry has made major improvements to its commercial systems through lessons learned in both extensive demonstration programmes and through applied engineering. Creation of testing standards and rating programmes have also contributed to consistently improved industry performance in quality control. Collector manufacturers learned many lessons about material compatibilities, the proper design for stress in collectors due to expansion and contraction, need for production quality control, corrosion, metal-to-metal connections in a thermal environment, and piping and flow

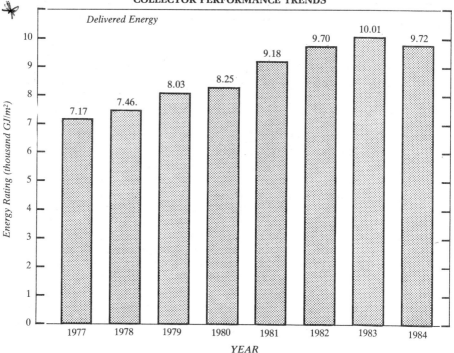

FIGURE 7
COLLECTOR PERFORMANCE TRENDS

Source: Florida Solar Energy Center.

characteristics of multi-collector arrays. Designers and installers of air-type systems had to learn the importance of using high-quality dampers and duct work to avoid 10% to 35% thermal losses in their systems. As industry shake-out has occurred, applications of these lessons have resulted in a technically stronger industry with higher quality solar collectors with increased performance and longer warranties for consumers.

Research has led to development of other collector technologies. Two of the most important for increased collector and system performance may be evacuated-tube collectors and low concentration fully stationary non-evacuated concentrators. These units are now capable of achieving an annual 50% conversion efficiency with output temperatures of 100°C. In addition, prototype flat-plate collectors constructed of light-weight polymer materials have been tested. These might offer lower material requirements and easier installation, but their reliability and durability have not been proven.

The results of extensive systems-oriented research have emphasized that problems existed in systems design and integration, particularly with regard to emergency-mode control and interfaces with conventional heating, ventilating and air conditioning systems. Many system failures are caused by control malfunctions or improper placement of sensors. Better design tools to assist the correct sizing of heat exchangers, expansion tanks, and storage strategies were needed and these have been developed and upgraded, a key accomplishment throughout the IEA countries.

System designers, project owners and manuacturers have all learned a crucial lesson about system failure. Improper field installation was a preponderant cause of problems. Designers and manufacturers have had to accept greater responsibility for proper system installation.

Another set of key accomplishments has been the improvement in quality, performance, reliability and durability of parts and components made by non-solar companies for the solar heating and cooling industry. The need for such improvements became evident in the earliest government-sponsored demonstration programmes.

Government research programmes have sponsored major improvements in electroplated black chrome selective surfaces for solar applications. Costs of such surfaces has dropped from over US $20/m^2 to US $8/m^2. At wide temperature differences between collectors and ambient air, these surfaces can nearly double collector output. Government and industry research has also developed thin and thick film selective paints that offer performance nearly equal to black chrome. Another government-sponsored advance, the hydrofluoric etching of ordinary glass, has increased transmissivity of glass surfaces by 2% to 3% and a 5% increase appears possible. When applied this process can result in a 5% to 10% improvement in collector efficiency at operating temperatures.

An innovative optical design technique that allows stationary collectors to capture the maximum possible radiation was developed by Dr. Roland Winston in the United States. When applied to evacuated-tube solar collectors, this non-imaging optical technique permits a 50% reduction in expensive evacuated tubes by use of reflective black surfaces with little loss in efficiency. The technique has been used to design marketable collectors by manufacturers in several countries. The latest research suggests placing the compound parabolic reflective surface inside evacuated tubes, with a prospect of efficient energy conversion at temperatures above 200°C.

Solar air conditioning systems have been demonstrated with absorption and Rankine-cycle chillers. Good systems have been built that show technical feasibility for small and medium sized buildings. However, they are not close to being cost-effective nor to saving very much energy. Desiccant-cooling technology was advanced to a commercially offered product by one United States company. Significant progress has been made by the research community in identifying and testing suitable materials for this type of system. Nevertheless, all solar cooling technologies still require substantial research and development before they can be commercialized.

Performance Testing

Substantial effort has been expended in nearly every IEA country to build and monitor active solar systems of all types. Documented performance has ranged from negative energy savings to systems that could practically supplant all fossil energy requirements for a particular end-use such as water heating.

Table 6 illustrates data gathered in the United States for a number of systems monitored in the United States National Solar Data Network (NSDN). The best commercial water heating systems delivered 587 kWh/m^2 of collector per year, while the lowest performing system in this category delivered 25 kWh/m^2 per year.

TABLE 6

ENERGY OUTPUT FROM NSDN (1982) SOLAR SYSTEMS

(kWh/m^2/year)

	Number Of Systems	Highest Output	Lowest Output	Average Output
HOT WATER, COMMERCIAL	21	587	25	350
SPACE HEATING	13	258	0	142
SPACE & WATER HEATING	23	381	22.2	153
COOLING	12	70	-20	22

The best commercial hot water systems operated with an annual solar conversion efficiency of 41%. For space heating it was 22%, space and water heating 25%, and cooling 7%.

In the United States, the Tennessee Valley Authority tested 25 commercially-available domestic hot water systems with five different freeze protection strategies. Results indicate energy delivery can vary almost a factor of two depending on the freeze-protection strategy, number of storage tanks, presence and effectiveness of heat exchangers, selective versus non-selective absorbers, single versus double glazing, plastic versus glass covers, and fluid circulation rates. The best systems delivered 1.95 kWh/m^2 per day, the lowest performers, 1 kWh/m^2 per day (at an average solar radiation level of 4 kWh/m^2 per day and ambient temperature of 15°C).

Monitoring of field-installed residential solar water heaters has shown wide variations in performance. In one sample of 41 systems, 10% provided an insignificant amount of energy, and 30% provided less than half the output of the best systems, which averaged 1.4 kWh/m^2 per day. Monitored commercial water-heating systems have shown similar variations in performance. The best systems have achieved system efficiencies of 35% to 40% and daily energy output of 1.4 kWh/m². These variations are due most to quality of installation, control and maintenance.

In summary, researchers have made major improvements in active solar technology, with contributions coming from every country with research, development and demonstrations in this field. The feasibility of the technology has already been demonstrated and more sophisticated design and analytical tools promise additional gains. Finally, besides scientific and engineering gains for these technologies, research on manufacturing has shown that increased serial production has lowered equipment costs signficantly. Research on manufacturing has shown that increased production has substantially improved the quality of the solar components and systems - durability, reliability and efficiency.

Commercial Status

The market for solar heating and cooling technology may be divided into four segments: residential, agricultural, commercial and industrial. The residential market for active solar technology is the most commercially advanced, often because solar systems compete against electricity. Systems for agricultural and commercial applications are found, but the market is not particularly well advanced. Market development in the industrial sector lags even further behind.

The size of the industry and annual production of solar collectors are two indicators of commercial development over the last decade. The number of manufacturers involved in the active solar industry reached a peak in the late 1970s, then declined. For example, in the United States, which has the largest developed market for active solar, the number of collector manufacturers rose from near 50 in 1974 to a peak of 349 in 1979. Then the conditions that had fuelled rapid growth of active solar technology sales began to change. Beginning in the 1980s, the start of a steady decline in oil and gas prices, economic recession, and gains in energy efficiency for conventional heating and cooling systems cut active solar market development. By 1984, the number of United States collector manufacturers had dropped to 224. An even more drastic reduction of 50% to 90% occurred in Denmark, Germany, Sweden and Switzerland. Analysts say that the loss of corporate interest was due to failure of active technology to drop rapidly enough in cost and improve sufficiently in reliability. Clearly, industry has greatly reduced expectations compared to the highly positive atmosphere prevalent in the 1970s.

In the United States, an additional severe decline in the number of manufacturers began in early 1986 when federal government tax incentives for solar systems expired. Sales compared to 1985 have been reported to be off from 65% to 80% and many companies have either closed, or entered new commercial fields.

In the 1970s, the commercialization of new active solar technologies began with the residential swimming pool market. In some countries with more favorable climates or significant government incentives, domestic water heating eventually overtook the swimming pool heating market, both in terms of square meters of collectors produced and income. In Europe, however, swimming pool systems continue to be a dominant commercial technology. For example, in Austria in 1985, 72% of solar collectors were used for pool heating compared to 26% for service hot water.

In countries with some market penetration, retrofitting existing structures rather than new construction accounts for as much as 80% of all active solar installations. In Japan, as many as four million batch-type solar water heaters had already been produced in the 1960s, of which 3 million are estimated to remain in use. In 1984, annual sales of this type of unit totaled 500 000 per year, whereas active solar water heater installations had grown to a level of approximately 50 000 per year. Only about 3 000 active solar systems existed in Japan in 1984 for other applications besides domestic water heating. Collector production peaked in Japan in 1984 at 1 316 541 m^2 for all types. The total in 1985 was down about 10% to 1 182 977 m^2.

In the United States, the only active solar technology market larger than Japan's, federal and state government incentives have stimulated and supported commercialization. In the best years, sales of domestic water heaters exceeded 100 000 units per year. The dollar volume of *all* active solar sales in 1983 amounted to US $700 to US $800 million.

Typical solar water heaters in the United States comprise 5 - 6 m^2 of collector area and cost the user from US $3 000 to US $5 000 installed. Marketing costs are estimated to range from US $1 000 to US $2 000 per system. Although sales are most common in regions with a high percentage of sunshine hours per month, high energy cost regions such as the northeast in the United States have experienced significant market penetration while subsidies were available.

Closed-loop and open-loop pumped systems dominated the residential market during its formative stages. In regions where freezing temperatures are not frequently encountered, batch type units (industry refers to them as integral collector storage-ICS-water heaters) and thermosiphon systems have captured 40% or more of the market.

In 1984, United States collector manufacturers reported annual shipments by market sector as follows:

(Square Meters of Collector)				
Residential	Commercial	Industrial	Other	Total
1 299 000	194 000	27 000	5 500	1 525 000

In Australia, a survey in 1983 indicated that 4% of the population own solar energy systems. However, in two areas where solar insolation is high, market penetrations have reached 20.7% and 37.2% of households. These water heaters typically use 4 m² of collector and cost A$1 500 installed. Studies indicated that annual sales of collectors increased in Australia by a factor of 20 from 1974 to 1981, then continued to increase from a cumulative total of 115 400 to 191 000 in 1983. For 1984, sales were 30 000 units; while in 1985 some market fall-off is evident since 15 000 of them were installed. It is now estimated that 250 000 households have such systems installed.

In Canada, economic incentives were given to solar manufacturers in order to lower the cost of delivered systems. At present, systems being installed with a new low flow rate concept and stratified storage are priced at CDN $2 300 for 5 m² of collector. In 1985 sales of 2 000 domestic water heaters and approximately 100 other kinds of active solar installations amounted to CDN $20 million, an increase of 20% compared to 1984.

Although space heating technology is proven (frequently in combination with water heating), only 5%-10% of collectors produced in IEA countries are used for this application. These systems have been sold in both sunny areas and in regions with high-cost energy, and usually require short-term energy storage. A market surge due to heavy industry marketing of tax credits occurred in the United States for simple panel air-type systems with no storage. Sales for these systems raised the percentage of air-collectors produced from under 10% to nearly 20% in 1984 and 1985.

Another indication of the growing commercial maturity of active solar technology has been the institution of consumer protection programmes. Manufacturers offer warranties on solar collectors for up to five years, and in some cases, ten years. Collector rating and certification programmes have been initiated in several countries, and there are also system performance testing programmes.

In summary, active solar technology for pool heating and domestic hot water is considered mature; however, sales in some countries have been heavily dependent on incentives, particularly for water heaters. In most IEA countries the penetration of solar water heaters compared to conventional heaters is minimal. There exist, nonetheless, certain regions in IEA member countries where solar water heating has gained a firm competitive position. A prime example is northwestern Australia where market penetration of over 30% has occurred. Furthermore, in certain countries, solar systems are competing with electricity which is still increasing in cost.

Active solar space heating technology is also proved, but has only made comparatively small inroads into the market.

The development of a market for solar cooling technology lags significantly behind that for heating systems. In the residential sector companies offered absorbtion systems, but they have generally been too expensive to be economic. Solar-desiccant cooling systems are a new innovation that one firm in the United States began to sell commercially, however, expiration of the residential tax credits has cut off this initiative. In the United States, the sale of solar cooling for both residential and commercial buildings occurs most frequently in the southwest. A few sales have been noted in other IEA countries in suitable climates.

Sale of commercial systems occurs in some countries, in large part for public swimming pools. For commercial and industrial systems that furnish process heat (laundries, food processing), companies in the United States have used lease financing and tax incentives to help open this market. Some of these projects have cost up to US $2 million, but very few exist, and firms generally report how difficult it is to sell solar energy in the commercial and industrial sector. Over the next three years, a 15% business energy investment tax credit should remain in effect in the United States, but to date its impact on market growth of commmercial and industrial systems has been marginal. Most commercial entities require payback of three years or less, and active solar technologies, even with incentives, have a difficult time meeting this criterion.

Market penetration of solar systems has been hampered primarily by the high "first cost" of the system. There is a high inverse correlation between market penetration and system cost. In 1984, the Swedish Council for Building Research reviewed active solar research in eight countries. Table 7 *(Solar Energy Research Outside Sweden)* presents the installed price per square meter for active solar domestic water heaters for single family homes.

TABLE 7
PRICES OF SINGLE FAMILY SOLAR DHW SYSTEMS
(US$1984 at 1984 exchange rates)

	US $/m^2
Australia	247 (Thermosiphon)
Canada	457
France	297-593
Japan	309
Italy	309
Netherlands	272
Sweden	346
United States	494-741

The large differences in price may be accounted for by design, materials, distribution and marketing costs, warranties, codes and standards, installation costs and above all exchange rates. The Netherlands' systems are among the lowest priced in Europe, and compare with systems in Australia. The figure given for Australia is for a thermosiphon unit: whereas the figures for the United States represent pumped closed-loop systems with more expense for controls, heat exchangers, larger storage and requirements to meet codes. United States manufacturers have also reported that marketing represents from 25% to 33% of installed costs for their systems.

V.   ENVIRONMENTAL CONSIDERATIONS

No significant environmental barriers exist that would prevent widescale adoption of active solar technologies. Solar systems are material-intensive, using copper, steel and aluminum, nickel and chrome alloys for absorbative coatings, glass and plastic (polychlorides and acrylics), insulation materials (isocyanurate is common) and heat transfer fluids (hydrocarbons, glycols, untreated and treated water).

Land use impacts pertain to larger systems for multifamily residential, commercial and industrial applications, and for seasonal storage systems. In the residential sector, only storage systems might require additional land since collector arrays are typically roof-mounted. Large active solar arrays for industrial process heat that were ground-mounted would have impact due to disruption of the land and area during construction.

Once operational, active solar systems require very little material or water. Where they are used, spent heat transfer fluids need to be disposed of carefully since many systems use anti-freeze with rust and micro-organism inhibitors and will have leached other substances during operation. Such disposal should occur every two to four years.

Safety hazards are minimal. In only a few rare instances have fires been caused by stagnation temperatures, misaligned concentrators or released coolant. Accidental pollution of potable water supplies is possible. Government standards frequently require double-wall heat exchangers if toxic heat-transfer materials are used. Finally, overheated panels can cause volatilization of materials in insulation, plastic components and seals and gaskets, although the amount of air pollution is generally minimal. There have only been a few reported instances where significant concentrations of outgassed products occurred in poorly-constructed air systems. There is very low risk from the mounting of collectors on buildings, and the risk is easily controlled through building codes.

In summary, findings of numerous studies indicate practically no environmental constraints relevant to the development of active solar heating and cooling technology. Health and safety hazards that exist are rarely life-threatening and are easily controlled by adequate, normal regulatory practices and proper design guidelines, installation procedures and disposal methods. Once systems are operational they have practically no environmental impact.

VI. TECHNOLOGY OUTLOOK

Technological Prospects

The IEA Member countries have made great progress in furthering active solar space conditioning and water heating technologies since 1974. These improvements have also benefitted the technologies used for swimming pool, commercial, agricultural and industrial applications. R&D has been devoted to materials, performance improvements, cost reduction, increased reliability and durability, and storage, including long-term seasonal storage concepts.

Research and development on active solar heating and cooling extended beyond improvement of existing hardware. A wider discipline of solar energy analysis was developed, and a structure of analytical methodologies and data collection put in place to support the technology research and development, and ultimately, commercialization. Hardware-oriented research sought to improve existing concepts through materials advances and systems design analysis. Widespread demonstration of technology was also a key objective in many IEA Member nations which is a result of policy emphasizing rapid commercialization. The combination of these initiatives has created a substantial knowledge base to support continued advancement of active solar technologies.

Current R&D in collection technology focuses on reducing material intensiveness of flat-plate collectors and on improving the performance of flat-plate non-tracking concentrating collectors. While the new polymer film collectors have achieved their cost and thermal performance goals, their long-term durability has yet to be established. New innovations such as silica aerogel glazings with three to five times the thermal resistance of glass may significantly improve collector performance while reducing weight. This and other developments of strong, durable ultraviolet stabilized films for collector glazings and polymer based absorbers, if successful, could reduce the mass of a typical flat-plate collector from 20 kg/m^2 to 3 kg/m^2. Such reductions decrease cost of materials and offer new economies in fabrication, distribution and installation. Other material advances in selective coatings should result in more cost-effective application and longer life in high-temperature collectors.

Large-scale modules designed for rapid installation in large arrays have greatly reduced the installed cost of collector fields for central heating plants for community and commercial applications. Sweden has developed 12 m^2 flat-plate collectors that two installers can mount at a rate of 100 m^2 per hour. When operating between 70°C and 90°C, collector performance is equivalent to commercial evacuated-tube collectors.

Stationary concentrating evacuated-tube collectors, with potential performance equal to industrial quality, large parabolic-trough collectors are under development. Such collectors will be preferred for small-scale applications because of their simplicity and low maintenance requirements. High efficiency at elevated temperatures make them ideal for advanced solar cooling technology. Once in mass production, this type of collector should cost no more than conventional units but should exceed their thermal output for higher temperature applications.

Energy storage is considered to hold the key to effective utilization of many renewable energy technologies. Current active solar space heating systems typically store energy for one or two sunless days. This limited storage capacity means that some available solar radiation cannot be used or stored. In this regard, long-term interseasonal storage options are under development and testing. Candidate long-term thermal storage techniques include: aquifers, in situ earth-heat exchangers, rock caverns, drilled rock, and underground water tanks or pits. In addition to increasing the useful output of solar systems, large storage systems facilitate integration of a variety of renewable and conventional energy sources. In Sweden, the cost of large-scale low-temperature energy storage is already as low as US $0.04/ kWh per cycle, based on experimental data.

Chemical energy storage offers all the advantages cited above, but is also transportable. Although realization of this option is a technical and economic challenge, the potential justifies continued research. Storage systems using chilled water or ice are already being introduced for commercial buildings. Ground coupled heat pump systems could enlarge the potential for active solar technologies to include regions with substantial heating and cooling seasons. Long-term storage for individual residences or small commercial users with current technology does not appear cost-effective.

Continued advance of microprocessor technology can be expected to benefit active solar energy systems. Up to now a major source of problems and failure, system controls will become more intelligent, automated and reliable. For example, smarter controllers can take advantage of high efficiency, low-cost control elements (e.g. brushless d.c. motors) to vary continuously liquid and air flows and increase system performance.

Efficiency of energy conversion equipment for cooling and dehumidification is expected to improve by at least a factor of two. Absorption cooling machines using regenerative, multistage processes are already in advanced stages of laboratory development. These devices could increase chiller coefficients of performance (COPs) from a current range of 0.7-0.8 to 1.3-1.5 or better. Further advances in solid desiccant materials and dehumidifier design should lead to cost-effective solar air conditioning systems. One approach may be use of hybrid systems combining desiccant dehumidification and vapor compression refrigeration.

When compared to field installations, performance monitoring has revealed that a factor of two improvement in output can be found in experimental systems maintained by experts. The difference is attributable to soundness of the design, the quality of the installation, and the extent of maintenance required and performed. The potential for increased performance of active solar energy systems, therefore, lies in system design improvements, better component integration, close attention to thermal loss reductions and changes in operating and control strategy. Those maintenance requirements that remain must also be manageable for *trained* site personnel.

Outlook

In the 1970s, IEA member countries emphasized development of active solar heating and cooling technology because of its many potential applications and the apparent simplicity of the systems. The potential of wide-scale application for active solar technology remains. The last decade has shown, however, that performance and durability involve more complexity than anticipated, and that substantial improvements in technology, performance and costs would be required for solar heating and cooling to begin to realize its potential. Looking ahead it is clear that excessive optimism generated in the mid-1970s has been tempered by technological and economic realities. Nonetheless, the work accomplished to date has amply proven that some active solar applications are cost competitive in certain regions of IEA member nations. Consequently, the general outlook for active solar heating and cooling is encouraging.

Many countries continue to apply lessons learned from the last decade to improve solar technology and its cost-competitiveness in their nations. As part of national energy R&D planning a number of IEA member countries have forecast solar heating and cooling contributions in the year 2000. Most of those that have reported their recent projections to the IEA expect an energy supply contribution in the order of 1% or less. In European countries, estimates reported to the IEA Secretariat range from 0.2% to 2%.

In Japan the expected contribution from active solar technologies to energy supply by the year 2000 is 1% to 2%. The Government estimates that solar water heaters are only cost-effective to the consumer with local or national government incentives and expects that installed costs must drop by a factor of two to be independent of government support. Thus, the Japanese Government continues to fund research, emphasizing industrial solar energy systems, and supports residential applications with loans and subsidies. Basic studies on materials and components, testing procedures, and systems design are still being pursued. Furthermore, the government attempts to encourage industry interest in this technology through its support.

In Australia, where satisfying demand for domestic hot water requires 2% of national energy supply, the government expects that solar water heating can ultimately supply 60% of the energy required for that end-use. Even though the contribution of active solar energy from domestic water heating and swimming pool heating is forecast to grow 8% per year, the government expects all solar energy technologies to contribute only 0.14% of energy supply in the year 2000. Officials consider high capital cost the major constraint to more rapid development of active solar technologies.

In the United States, the reference case for the National Energy Policy Plan Projections to the year 2010 (United States Department of Energy, December 1985) projects an active solar contribution of 0.12% to primary energy demand in the year 2000. The level of confidence in this projection is low because many factors besides system-related performance and economics have bearing on the market penetration rate. These other factors include the cost of fuel, interest rates, the level of federal and state incentives, rate of R&D innovation, constraints due to the large level of existing building stock, and success of other competing solar, conventional and conservation technologies.

In conclusion, looking towards the year 2000, the outlook for mature solar swimming pool technology is positive with markets expanding based on economic competitiveness. Domestic solar water heating technology is mature: however, new technological developments are forecast, and *required,* to widen the economic competitiveness to many countries rather than in regions of a few countries. Solar space heating technology is an area where important gains are expected. Central heating plants with seasonal storage could begin to be commercialized in countries in the northern-most latitudes within the next five years. Sweden, in particular, appears ready to emphasize this technology. Active solar technology for commercial, industrial and agricultural applications will develop more slowly. However, many developing technological innovations pertinent to the early markets for residential systems should improve the cost competitiveness of active solar systems in these other sectors. Then the issue will be comparative economics.

VII. MAJOR FINDINGS

Over the last decade, major advances have been made in active solar technologies for all basic applications: swimming pool heating, domestic water heating, space heating, space cooling, heat for commercial, industrial and agricultural processes. Through this progress, technology for solar swimming pool heating and domestic water heating has reached a relatively high level of maturity. Efficiency of commercially-produced flat-plate collectors has increased by 30% since 1977, and new collector concepts such as the high performance evacuated-tube collectors have been developed, tested and launched commercially.

Currently, the largest number of active solar installations are residential water heaters, and the greatest market penetration is in the residential sector for existing single-family dwellings. The largest number of sales occur in the United States, followed by Japan and Australia. After rapid growth through the 1970s, sales in these three countries have remained level for the past four years. In 1986 in the United States, a significant drop in sales is occurring due to expiration of government incentives for residential solar energy systems. It is evident that progress in market development in most countries has been relatively dependent on government incentives. Some regions in IEA Member countries exist, however, where solar water heaters have penetrated a significant level of the market.

Real production costs of active solar energy systems have fallen steadily and further cost reductions are expected and necessary for markets to expand. System and installation costs vary widely from one IEA country to another. Efforts to reduce ratios of installation cost to system cost as well as reduce solar component and system costs should continue. Evacuated-tube collectors require mass production economies to reduce costs further.

Research and experience indicate practically no environmental constraints exist that could impede market development of active solar heating and cooling technology.

Although expectations of the 1970s for rapid market development of active solar technologies in IEA member countries have not been realized, revised modest goals of 1% or 2% energy contribution in the year 2000 are sufficient to justify national research programmes continuing to pursue further development of these systems.

R&D on collection technology is now focusing on reduction of material intensiveness of flat-plate collectors and improvement in the performance of flat-plate and non-tracking concentrating collectors. Prototype polymer film collectors indicate that potential for weight and cost reductions may come from application of new materials. Efficiency of solar cooling and dehumidification equipment is expected to improve by at least a factor of two by the year 2000. In addition, long-term seasonal storage appears to be approaching economic viability in northern countries and some countries will attempt to push its commercialization. R&D must concentrate on increasing component lifetimes, lowering costs, and improving reliability and durability. Of critical importance is R&D on systems. This should identify the most efficient and economical way to use these improved solar components in conjunction with the conventional components in heating, cooling and hot water systems.

VIII. BIBLIOGRAPHY

1.   Kreider, J.F. and Kreith, F., *Solar Energy Handbook,* McGraw Hill, 1981, pp.(1-16) - (1-38).

2. Duffie, J.A. and Beckman, W.A., *Solar Energy Thermal Processes,* John Wiley and Sons, 1974, pp. 1-4.

3. Franta, G., et al., *Solar Design Workshop,* SERI/SP-62-308, June 1981, pp.(3-1) - (3-3).

4. "National Solar Energy Heating and Cooling Programme Overviews", presented at the International Solar Energy Society Solar World Forum, Brighton, UK, August 1981.

5. *"Energy Policy and Programmes of IEA Countries",* International Energy Agency 1983 Review.

6. "US Mean Daily Solar Radiation Annual", International Energy Agency, January, 1984.

7. "Review of Renewable Sources of Energy", International Energy Agency, January, 1984.

8. "Renewable Energy Strategy Development Report", Vol. 1 - Active Solar Heating and Cooling, December 1983, SIAC/Meridian Corporation.

9. "Survey and Review of National R&D Plans", IEA Task II Subtask B, August 1984.

10. Brown, Kenneth C., *Handbook of Energy Technology and Economics,* John Wiley and Sons, New York,1983, p. 618.

11. *Annual Renewable Energy Technology Review,* Renewable Energy Institute, Washington, D.C., 1986, p. 219

12. Charles Bankston, Consultant, Task VIII, International Energy Agency.

13. *National Energy Policy Plan Projections to the Year 2010,* United States Department of Energy, Washington, D.C., December 1985.

14. *Energy Innovations: Development and Status of the Renewable Energy Industries 1985, Volume 1,* Solar Energy Industries Association, Washington D.C., 1986.

15. Rabl, Ari, *Active Solar Collectors and Their Applications,* Oxford University Press, 1985

16. Montgomery, Richard H. and Budnick, Jim, *The Solar Decision Book, A Guide for Heating your Home with Solar Energy,* John Wiley and Sons, 1978.

C. PASSIVE SOLAR TECHNOLOGIES

I.   INTRODUCTION

Solar energy which reaches the surface of the earth can be used directly to meet some of the heating and lighting needs of buildings. In addition, solar-induced climate effects such as wind and temperature differences can be used to cool buildings. Passive solar technologies use these phenomena to condition buildings using little or no mechanical assistance. This substantially reduces the amount of energy needed to heat, light and cool buildings.

Basic passive solar concepts have been used throughout history to increase the comfort of shelters for humans. Only within the past few decades have the techniques for refining these concepts been more intensely investigated. Recent increases in publicly and privately-funded passive solar energy R&D represent the first steps towards a scientific understanding of the application and potential impact of these technologies.

Passive solar research has resulted in a variety of design and analysis tools which are now available to guide architects and engineers in the proper application of these technologies. These tools are increasingly reliable and easy to obtain and use. In addition, passive solar architecture has received favorable coverage by both industry and popular press in many IEA countries, creating public awareness and some practitioner interest. Since passive solar design includes daylighting (the natural light in buildings) occupants of passive solar buildings have tended to respond favorably to this. The simplicity of passive solar technology has made it both accessible and cost-effective. Recent studies have shown that well-designed residential and non-residential passive solar applications generally cost little or no more than comparable conventional buildings. For these reasons, the use of passive solar technology can be expected to grow steadily as an economical source of heating, daylighting, and cooling energy for buildings.

II - III. TECHNOLOGY DESCRIPTION AND APPLICATIONS

Passive solar systems can be used for heating, cooling and lighting of both residential and non-residential buildings.

Passive Heating Technologies

Passive solar heating involves the transmission of solar radiation through a protective glazing layer(s) on the south side of a building (north side in the southern hemisphere) into a building space where it is absorbed and stored by thick masonry walls and floors or water-filled containers. These materials are often referred to as "thermal mass," which is needed to store the solar energy for subsequent release into the space as heat.

While solar energy transmission through a single sheet of glass is high, so are the conductive and radiative heat losses out through the glass. To achieve a net winter energy gain in cold climates from south-facing glazing, these thermal losses must be reduced. This is generally done either by adding one or two additional glazing layers to the single-glazed window, by using a glazing with enhanced insulating properties, or by using some form of movable insulation to cover the window at night.

Since the portion of a building's heating needs that can be met by solar energy increases as the area of south facing glazing increases, additional mass must be used to reduce interior temperature swings and delay the release of solar energy into occupied spaces. While the mass that is directly illuminated by the incident energy is the most effective for energy storage, long-wave radiation exchanges and convective air currents in the solar heated rooms allow non-illuminated mass to provide effective energy storage.

In passive solar systems solar heat is distributed from the collection and storage points to different areas of the house by the three natural heat transfer modes of conduction, convection and radiation. In a few applications, fans, ducts and blowers are also used to help with the distribution of heat.

Passive solar heating systems can be divided into four generic categories. These are based primarily on the spatial relationship of three key parts of any passive solar heating system: aperture (or collector), mass, and heated space.

Direct Gain Systems: Direct gain systems (Figure 8) utilize windows or skylights on the building exterior to allow solar radiation to directly enter zones to be heated. If the building is constructed of lightweight materials, mass may be used to be added to the building interior to increase the heat storage capacity.

Sunspace: In sunspaces, glazing allows solar radiation to enter an accessible, usable but isolated space on the south side of the building (Figure 9). Heating usually enters the building via vents and/or ducting systems. Unlike a direct gain space, a sunspace is minimally conditioned. Ventilation and shading are sometimes required in the summer to limit overheating, and some heating may be required for freeze protection in the winter. However, if the sunspace itself is heated and cooled by other means to normal comfort levels with auxiliary energy, the buffering characteristic of this isolated-gain passive solar system is lost. In this case, the total energy consumption of the building and sunspace may exceed that of the building alone. Sunspaces are currently quite popular because of their spacial and architectural amenities.

Indirect Gain: Indirect gain systems are characterized by the placement of glazing in front of and quite close to a solid masonry wall or other building element that has a high thermal storage capacity. These systems are designed to capture and store a large fraction of the incident radiation for subsequent release into the adjoining occupied space. When a solid masonry wall is used for thermal energy storage the indirect gain system is commonly called a Trombe Wall (Figure 10). The inclusion of vent holes through the heated wall permits some convection to take place, as a means of providing heat to the building during the day. A variation on the Trombe Wall is the Water Wall which employs water-filled containers in place of a masonry wall. Tall fibreglass tubes are often used (Figure 11).

Convection Systems: Passive solar convection (or thermosiphon) systems take advantage of the buoyancy effects of heated fluids for the convective transfer of solar heat into building spaces or storage. Since these systems often use mechanical assistance (pumps, ventilators), they are often called hybrid systems. Air or water is the most common heat transfer medium, with other gases or liquids being tested in advanced systems. In early applications of this technique, solar collectors should be placed below the thermal storage so that the heated air or water could move by thermal convection into storage. These systems could be used for space heating or water heating. Passive solar domestic hot water systems that place the storage tank on the roof above the collectors, have been widely used in countries such as Japan and Australia and are now gaining an increasing portion of what used to be primarily an active solar market in other countries as well.

A currently popular convection system uses air heating flat-plate collectors mounted on the outside of an exterior south-facing wall in which vents located at the top and bottom of the panel allow hot air to flow directly into the space to be heated, replaced at the bottom of the panel by cooler interior air. (Figure 12)

Today's passive solar heating systems incorporating any of the concepts described above can typically provide 30%-70% of residential heating requirements, depending upon the size of the passive solar system, the level of conservation in the building envelope, and the local climate. The high end of this range often requires the use of specialized components at an additional first cost. These performance results are typical for single-family residences, and some small commercial and institutional buildings.

These passive solar heating systems have also been applied very successfully to numerous non-residential buildings. However, experiences based on residential applications do not always apply because non-residential buildings have very different thermal requirements due to building use patterns, internal heat generation, and interactions with complex space conditioning equipment.

The rehabilitation and retrofit of existing buildings can often be done with passive solar heating systems. Sunspaces are especially useful in this regard because they can be added to existing buildings without major changes in that structure. While passive solar retrofits frequently encounter problems related to solar access, orientation, heat distribution, and thermal storage, these can often be overcome by ingenuity and careful design.

FIGURE 8
DIRECT GAIN

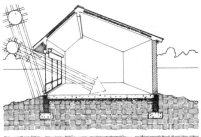

FIGURE 9
ATTACHED SUNSPACE
(Isolated Gain)

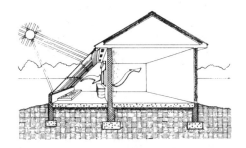

FIGURE 10
TROMBE WALL
(Indirect Gain)

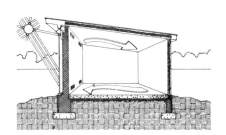

FIGURE 11
WATER WALL
(Isolated Gain)

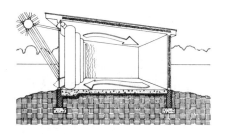

FIGURE 12
THERMOSIPHON AIR PANEL

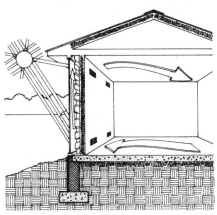

Source: (Fig 8-12): United States Department of Energy

Passive Cooling Technologies

Passive cooling technologies take advantage of the convective and radiative cooling of the earth at night to allow heat absorbed in the building during the day to be reduced without the use of mechanical equipment. The following four generic passive solar cooling technologies have been identified based on the heat transfer mechanism and the cooling resources they employ.

Ventilative Cooling: Ventilative cooling utilizes cooler exterior temperatures, usually at night, in regions with high diurnal ambient temperature swings, to cool the interior of buildings. In this application, a building is supplied with exterior air throughout the night either passively or using the heating, ventilating and air conditioning (HVAC) distribution system. The building mass is cooled and then absorbs heat during the next day, reducing the need for mechanical cooling. Ventilative cooling is particularily suited for dry climates with low night-time temperatures and for situations where building use produces cooling loads during the spring, winter, and fall. The technology is limited in areas where ambient relative humidity levels are higher than the comfort zone.

Evaporative Cooling: Energy is required to evaporate water. By properly using water to draw this required energy from the building, a cooling affect is produced. In regions where the air is dry enough to absorb a significant amount of water and where there is ample water supply, evaporative cooling systems can be used. Mechanical evaporative coolers using a blower and wet air filter pads are popular in many arid areas and are often considered to be hybrid solar systems because they lack a compressor. A less common application uses a sprayed or flooded roof surface for cooling. This technique is effective, but limited by the roof surface-to-building volume ratio.

Earth Contact: The cool and stable temperature of the earth several feet below the surface has long been recognized as a way to keep a building cool. In order to cool a building using the earth as a heat sink, energy must be transferred from the building to the earth. The method is most effective in regions with cold winters and hot summers. The most common way is by conduction through the underground walls of the building. However, in climates with short winters and very warm summers, the ground may become warm long before summer is over and provide no further cooling benefit. Despite the climate, earth used as a heat sink tends to rapidly approach the temperature of the heat source. This is a different and less useful technique than the use of earth as a buffer.

Radiative Cooling: This approach makes use of the radiation to the sky as a heat sink to remove heat from building surfaces. While it has seldom been applied due to a lack of building skin materials with appropriate radiative characteristics, some success has been achieved using a roof covered with a layer of water. Radiative cooling is also limited by surface-to-volume ratios, high ambient night-time humidity and cloud cover.

Daylighting Technologies

Solar radiation in the visible spectrum illuminates building surfaces to levels far above those required by building occupants. This daylight represents an obvious but under-utilized solar energy source. Careful building design, often using lightshelves or baffles, can produce comfortable levels of even interior lighting without glare. Electrical energy savings result from the reduction or elimination of artificial lighting, often through automatic dimming or switching controls. In addition, cooling loads due to heat given off by electric lights are reduced, which may allow the HVAC system to be downsized and operate less often.

Two basic categories of daylighting systems can be identified depending on the location of the spaces requiring daylighting.

Perimeter Daylighting: Perimeter daylighting uses glass in various locations, such as normal windows, high windows, and clerestories, to allow light to enter the perimeter rooms of a building. Baffles or lightshelves are then used to distribute light evenly over the workspace.

Core Daylighting: Core daylighting systems use design features such as roof apertures, atria, light wells, or light guides to bring daylight to the interior zones of a building. Daylighting in using skylights and roof apertures known as roof monitors offer energy cost savings in the top floor of buildings but are of no use for lower floors. Figure 13 shows a diagram of a roof aperture design which was successfully used to illuminate the core area of a 520 m^2 in a one-story public library building. While atria and lightwells are being used in multi-story buildings, their performance is not well understood. Optical systems which use tracking heliostats and light guides to bring daylight deep within a building are also being developed.

Research and experience indicate that daylighting offers dramatic reductions in energy costs in buildings with the high lighting and cooling loads which are particularly characteristic of commercial and public buildings. Daylit spaces are also perceived as architectural amenities and are popular with building occupants. Daylighting systems can be incorporated into retrofit projects as well as new buildings. Bringing daylight into interior building space still represents a difficult challenge.

IV. STATUS AND ACCOMPLISHMENTS

Status in 1974

Only a few buildings existed in 1974 that had been designed specifically to use passive solar concepts. Isolated examples of early, pioneering efforts could be found in France, the United Kingdom and the United States but these had no effect on the building industry. Work done in the early 1970s consisted primarily of passively-heated residences built by individuals with a personal interest in alternatives to conventional energy technologies.

FIGURE 13
ROOF APERTURES

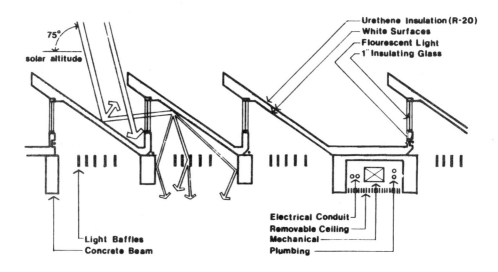

Source: United States Department of Energy

Although these early passive solar homes often performed well, they were usually elementary attempts at capturing solar energy for heating using large amounts of both glass and mass. There was a lack of any practical means for refining or predicting the impact of design changes on building energy performance. Practitioners relied primarily on an intuitive sense that had a very limited basis in experience. Subsequent national research and development programmes within the IEA countries sought to change this situation by providing a sound scientific basis for passive solar design.

Accomplishments

From the mid 1970s to early 1980s, many IEA countries established passive solar research programmes. Most initial passive solar research programmes emphasized residential heating because this was the least complex application. As a result of continued research, testing, and analysis over the past decade, heating applications of passive solar energy in residential buildings are well understood. The research attempted to better understand the performance of passive solar buildings, to quantify the effects of changes in the design and the effect of mass, and to determine the energy savings achieved in passive solar heated residences. In addition, efforts were also directed to developing guidance for architects and builders on the effective design of passive solar homes.

A similar focus was given to the IEA collaborative programme, Passive and Hybrid Solar Low Energy Buildings (Task VIII of the IEA Solar Heating and Cooling Programme) which was initiated in 1981. This programme, in which 14 countries participate, has conducted major activities in computer simulation, performance measurement, design tools/guidelines and building design/monitoring.

Most national programmes utilized test cells and monitored buildings to provide data on the performance of the various types of passive solar heating technologies. Algorithms were developed to define the basic relationships between solar radiation, mass, temperature, and heat transfer mechanisms and were used to write early computer codes for predicting passive solar system performance. Data from test cells and buildings were also used to validate the computer codes.

These codes and algorithms have been improved and refined over the years to provide what is now considered an adequate understanding of basic passive solar heating systems in single-family residential buildings. This understanding has been demonstrated and tested in over 200 monitored passive solar residences located primarily in Australia, Canada, Germany, Italy, and the United States.

A logical product of this research has been the development of design tools and guidelines that allow architects, engineers, and builders to use these passive solar principles in practice. The development of design tools and builders' guidelines has resulted in some use of this technology by the building industry. The advent of the microcomputer as standard office equipment has had a strong influence on the direction of this development. A 1985 survey of existing passive solar design tools by IEA Task VIII included over 230 design tools, the vast majority of them using micro-computers for modelling passive solar residential heating systems. These tools are typically based on algorithms and modelling techniques developed at the mainframe level, but use condensed weather data, reduced time steps, and/or mathematical correlations to dramatically reduce the run time and increase the ease of use. Frequently, these design tools are compared to mainframe simulation programmes to measure their accuracy. Microcomputer design tools are increasingly accurate, easy to use, inexpensive and have been a key factor in the increased use of passive solar heating by the residential building industry.

A similar sequence of events can be traced in the development of passive solar technologies for use in non-residential buildings: However, there is still much work needed to reach the current level of understanding of residential systems. These buildings are far more complex than residential buildings and therefore require a different approach to both the research and application of passive solar technologies. To a far greater extent than houses, non-residential building loads are influenced by occupancy levels and schedules as well as by electric lighting requirements. Often, large commercial buildings can experience cooling and heating loads simultaneously. Because of this, the use of test cells has been only partially successful for understanding the fundamental heat and mass transfer associated with the use of passive solar energy in these complex building systems.

Progress has been made in the development of large mainframe computer simulation programmes that can handle the complexities necessary to describe the relationships between the major sub-systems in non-residential buildings. Several non-residential buildings designed using these large computer programmes have been constructed in IEA countries and their performance is being monitored. Results to date indicate that for the less complex buildings these predictive codes allow a designer to arrive at a successful energy solution. In addition, some microcomputer tools are now available for use in passive solar non-residential design. However, further refinements and validation of in both the detailed simulation programmes and simplified design tools are still needed.

Daylighting, especially in combination with passive solar heating and cooling, currently has the greatest potential as a passive solar application in non-residential buildings. Carefully designed buildings have demonstrated that savings can occur, not only in lighting energy costs, but also by somewhat reduced cooling loads associated with the reduced electric lighting. There is also potential for further cost savings by reducing peak electricity demand.

In the early 1980s, the limitations of the building materials that were being used and the energy distribution methods became apparent. Initial focus was on developing phase change storage materials and reducing heat loss through windows. Products have been developed that have improved passive solar building performance, and there are a number of new, promising materials that will soon be available to the building industry.

Perhaps the most important result of materials research in the past decade has been the development and introduction of low-emittance glazing. This coating applied to glass greatly improves the performance of the glass as a thermal barrier by reducing the amount of heat emitted from the warmer inside glazing surface (Figure 14). Low-emittance glass is manufactured in several IEA countries and is now available and incorporated into a number of growing commercial window units. The use of this glazing can improve the performance of both passive solar heating and cooling applications by reducing thermal losses from the heated space in the winter and reducing thermal gains from the outside in the summer.

Two promising advanced convection systems are now being researched. One such convective system is based on the use of thermal diodes, a evice which integrates collection and storage into a single, sealed, close -loop unit. These can be used to create dynamic sections of exterioi walls, bringing heat into building spaces. A variation of this system is also under study that uses the vapor phase of freon refrigerant to move heat to remote building spaces.

FIGURE 14
CROSS SECTION OF LOW-EMISSIVITY GLAZING

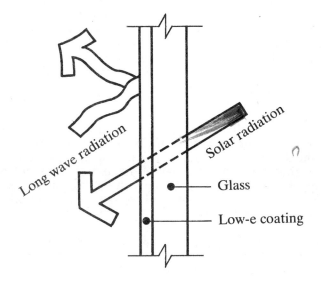

Long wave radiation

Solar radiation

Glass

Low-e coating

Source: United States Department of Energy

Commercial Status

The commercial status of passive solar technologies is inherently difficult to assess. Unlike active solar systems that are hardware-based, the most cost-effective applications of passive solar energy are those that require no specialized components or building systems. They are based primarily on design, not products, and consequently market penetration is difficult to quantify.

There are a variety of products available, however, that are specifically intended for use in passive solar design, usually representing high performance alternatives to standard building components. Phase change thermal storage, thermally-efficient windows, and movable insulation products exist for use in passive solar design. These products are relatively new, but are slowly gaining popularity with designers, builders, and homeowners. This is indicative of the growing use of passive solar technology by the building industry.

These products are not necessary for good passive solar design. However, they can increase system efficiency beyond the 20% to 30% range that is possible with standard building materials or increase the performance of a marginal application, as in the case of difficult solar retrofits. They also add to the cost of these systems.

Passive solar homes represent the current focus of the passive solar industry. The number of architects, builders, and home buyers that are aware of passive solar design as an option has grown dramatically over the last decade. In the United States well over 200 000 such homes have been built. It is estimated that about 50% of all recent passive solar homes are built for general purchase, indicating the degree to which passive solar design has entered the mainstream of the housing market. Home builders have emphasized passive solar design in marketing, and there are a number who successfully specialise in passive solar homes and housing developments.

Sunspaces, and especially sunspace kits, are currently the most popular passive solar system for new and retrofit residential applications. United States manufacturers alone estimate the market to be in the US $50-300 million range annually, with over half of the units sold intended to supply space heating. This popularity is partially due to the amenities offered by sunspaces in addition to their potential for reducing conventional energy use in new and existing homes.

The non-residential building sector has been neglected primarily because of a lack of available research results. The key to achieving good overall energy performance is the integration of daylighting, passive heating and natural cooling into the overall design. The use of daylighting in this sector has increased in some IEA countries. Although represented in only a small fraction of new non-residential construction, daylighting has met with positive user response and has been the cause of reported improvements in productivity by building managers. The use of passive solar design should increase as more design tools and guidelines are available for architects, engineers, and builders. Examples of the cost-effective applications of daylighting are increasing but have limited influence in building sectors where typically utility costs are passed on to tenants.

Daylighting has received enough attention to merit product lines and even businesses directed specifically at providing dimming and switching hardware for reducing electric lighting levels in response to daylight contribution. Ballast, lamp, and fixture compatibility is being addressed, promising an increasing selection of hardware options for use in daylit buildings.

Three factors have contributed to the cost-effective and growing application of passive solar technologies:

- Research done in many IEA countries has enabled architects, engineers, and builders to feel confident about applying passive solar concepts for a significant range of building types and climates, and provided direction in the form of design tools, guidelines and demonstrations.

- Specialized building components are seldom required; apertures and thermal mass are standard, functional building elements that are simply rearranged for optimal benefits.

- Design and performance experience has established the importance of addressing several building energy issues with a single component. Internal building mass can store direct gain heat and complete a night ventilation cooling system, both being driven by operable south windows. In general, this allows heating, cooling, and lighting energy to be obtained from one group of building components, increasing potential for economical applications of passive solar technology.

V. ENVIRONMENTAL CONSIDERATIONS

Passive solar technologies rely on energy that is incident on the building surface. Movement and conversion of this source of energy is accomplished with minimal mechanical assistance. The environment is not really changed, it is only used more effectively. As a consequence, passive solar is probably the least damaging energy technology.

Although indoor air pollution is sometimes labeled as an outcome of passive solar design, it is actually a building conservation issue. As a matter of fact, passive solar buildings are generally not designed as tightly as "super-insulated buildings" and therefore adequate air change rates normally occur. Air changes can also be designed into the passive solar system with relative ease to obtain excellent energy performance, perhaps exceeding that of typical superinsulated buildings.

VI. TECHNOLOGY OUTLOOK

Work Underway and Technical Prospects

Based on the research currently underway, the most important advances for passive solar technology are expected to be in the areas of high performance building materials and integrated building systems. Advances are occuring primarily in the areas of transparent insulation, variable transmittance glazings, and thermal storage materials. As these are refined, there will be a need for understanding their effect on related building systems and energy performance. The eventual application of these materials by the building industry will depend on the availability of sound technical information as well as purchaser demand. The information will result from computer simulations and data collected from experimental applications of these promising building components.

The percentage of the heating load of buildings that can be met by passive solar technology is presently limited primarily by the thermal losses from windows. Insulating glazing materials can increase options for the use of passive solar design in buildings by reducing the thermal difference between insulated opaque walls and solar apertures. The performance of all passive solar heating systems can be expected to improve as a result of these materials. In addition, non-south facing surfaces will be able to be used as glazed collectors more effectively to make a net positive contribution to building heating requirements. This could expand the potential number of new and retrofit applications, increase the performance of basic passive solar applications, and extend the feasibility of passive solar heating into climates with marginal solar resources. Silica aerogels, evacuated and gas filled windows, and high performance, low-emittance coatings are promising examples of current research in this area.

The ability to vary the transmittance of glazing materials addresses a second important issue regarding the use of glazing in passive solar design. Coatings for "switchable glazing" are now being developed that can be controlled electrically or thermally to reduce transmittance as desired to control solar gains (Figure 15). While still in early phases of development, these materials will allow glazing to be used for maximum benefit without problems of overheating and glare. Non-residential buildings will benefit directly from these glazings, especially in cases where lighting and cooling represent major, potentially simultaneous load problems, or where dynamic occupancy schedules produce both heating and cooling loads at different times during the same day. In general, the prospects for handling the dynamic nature of non-residential building loads will increase as a result of current research in this area.

FIGURE 15
CROSS SECTION OF AN ELECTROCHROMATIC WINDOW
(Switchable Glazing)

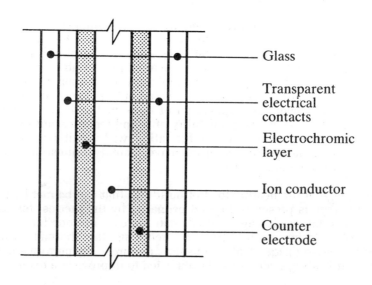

Glass

Transparent electrical contacts

Electrochromic layer

Ion conductor

Counter electrode

The performance of passive solar heating systems generally improves as the amount of south-facing glazing is increased, provided there is adequate mass to absorb the energy. Solid - solid phase change materials are under development that could significantly increase the amount of energy that can be stored in a given volume of material. In addition to improving the characteristics and durability of phase-change storage materials, new ways of using these materials are being developed. The experimental impregnation of common building materials, most notably gypsum wallboard, with solid state phase change materials has been successful enough to interest private sector manufacturers. The eventual availability of these products offers opportunities for both new and retrofit solar applications. The ability to add thermal storage elements that do not take up space or require structural consideration would enable designers to further integrate passive solar elements into buildings having a conventional appearance. Performance improvements would not be dependent on high levels of mass, and overheating problems which have occurred in some early passive solar designs could be easily corrected.

Passive solar cooling research underway in a few IEA countries offers opportunities for addressing a number of passive solar performance issues. The problems of overheating in passive solar residential systems have been addressed by those IEA countries where passive solar heating is accepted and used. Research is shifting to strategies for use in climates that are cooling load-dominated. Both evaporative and radiative cooling are being investigated. Night ventilative cooling is receiving attention as a promising means for addressing non-residential cooling problems and should be applicable to a variety of building types in many climates. The effectiveness of this technique will be enhanced by the development of building materials with increased thermal storage capacities.

Because of the greater complexity of non-residential buildings, there remain significant technical issues yet to be resolved related to the use of solar energy to meet their heating, cooling, and lighting needs. Current daylighting research considers the effect of natural lighting on the whole building, not just lighting energy use. In particular, balancing the application of daylight with heating and, especially, cooling loads, represents a major challenge in optimizing non-residential strategies. Advanced simulation programmes have been developed thereby allowing these concepts to be analyzed and refined. However, the availability of performance data from actual buildings is limited. Performance and cost data has been promising, indicating that careful, cost-effective design using daylighting will not exacerbate, and can even reduce, cooling energy requirements. There is a need for simple design tools for non-residential buildings, a need which is complicated by the wide variety of building types and use patterns in this building sector. Task XI of the IEA Solar Heating and Cooling Programme has recently been initiated to address some of the passive solar non-residential buildings issues.

Daylighting applications are being investigated that address the problem of bringing light into building interiors which presents the greatest challenge for maximizing the use of natural light in buildings. Various light guides and mirrored systems offer some promise for interior daylighting. Most of the advanced concepts and materials being investigated will depend on performance simulation and monitoring to direct their proper use in building design. The IEA countries have among them much experience in algorithm development and in gathering useful performance data. Modelling techniques now being used for non-residential buildings approach the complexity required to fully understand these new, dynamic components. Their impact on passive solar building performance will ultimately depend on how well they are integrated with existing building systems. This understanding will only be developed through directed research and development programmes based on specific experience with passive solar building technologies. Continuing research and design tool development offer potential for major energy contributions in buildings, especially in the light of recent developments in IEA countries.

Outlook

Future contributions from passive solar technologies are likely to maintain or exceed the growth that has occurred within the past decade. Much of the applied research has been completed, and innovative, advanced technologies hold substantial promise for improved passive solar contributions at reasonable costs. Passive solar heating is already well established as an option in residential design, primarily as a result of research performed by many IEA countries. These applications should become more widespread as existing design tools and guidelines are used by more practitioners. One can expect the more cost-effective and well-integrated passive solar heating technologies to gradually become accepted as good practice throughout the building industry, for new as well as existing buildings.

Current advanced materials research appears to indicate that a new level of passive solar heating applications will emerge to replace what is now considered state-of-the-art. These will be distinguished by energy contributions of well over 50% without the constraints of high mass, orientation, and night insulation. This will be due to high-performance glazings that provide net heating season contributions regardless of orientation and building materials that can spread thermal storage throughout a space for optimum benefit. Good passive solar performance will be easier to obtain without significant modifications to standard residential practice. For this reason, design tools and guidelines will become simpler, and more like standard building industry reference materials. In addition, industry acceptance should increase as solar buildings become visually indistinguishable from conventional designs, perhaps even losing the label "solar".

The use of daylighting in non-residential buildings will probably become commonplace as both design tools and control hardware become more available. Positive occupant response to this technology, combined with economic benefits, indicates that currently feasible options for perimeter and roof aperture daylighting will grow in popularity. Current research in the area of core daylighting, especially with light wells and atria, will soon yield the kind of technical understanding and design tools required for the use of these technologies. More advanced and experimental techniques for moving light, such as fibre optics and light guides, will make it possible to use natural light throughout building interiors.

The ability to vary window transmittance values will become a major factor in the evolution of passive solar non-residential buildings. The contribution of solar energy to building heating, cooling, and lighting loads will be tailored to fit the particular daily or seasonal building energy requirements.

These advances in glazing, thermal storage, and daylighting will result in buildings which can obtain most of their energy requirements from solar energy. The buildings will be fundamentally different in concept than today's buildings, since these materials, when integrated with building systems, will allow a dynamic response to changing patterns of building use and climate.

Conventional mechanical heating, cooling, and lighting systems, will no longer be the primary regulators of thermal and visual conditions. As a result, these systems will be replaced by ones which primarily facilitate and control the collection, storage, and distribution of heat and light. Designed to maximize the use of solar energy, the architectural building envelope of the future will respond dynamically to climate and to building use patterns, no longer functioning as a static barrier between the interior and outside environments.

VII. MAJOR FINDINGS

In most IEA countries, passive solar heating represents a proven technology, and is increasingly used by the residential building industry in new and retrofit applications.

A large performance data base has been established, including over 200 monitored passive solar buildings in many IEA countries; this has resulted in an improved scientific understanding of the application of passive solar heating in residences.

A major result of research and development in IEA countries has been the variety of design tools and builder's guidelines that have enabled the use of these techniques by the building industry. Many of the design tools still need to be validated, however.

Although cooling applications lag behind heating in terms of research and use, techniques are being investigated in some IEA countries that will increase the contribution of passive solar cooling to building energy requirements.

Data from innovative buildings in many IEA countries indicate that passive solar design can simultaneously address the need for heating, cooling, and lighting energy for non-residential buildings. Passive cooling applications are presently under investigation and are being used in residences on a limited scale. Since passive cooling techniques are very dependent on specific local climate conditions, their usefulness appears at this time to be more limited than that of residential passive solar heating.

Research on passive solar technologies for heating, cooling, and lighting of non-residential buildings was initiated only recently and it has not progressed as much as residential applications. However, most non-residential buildings where basic passive solar techniques have been used show very positive results. Passive solar heating of non-residential buildings has also been quite successful, supplying one half of the heating requirements in many cases. Because heat generated by people, lights, and equipment can greatly reduce the need for heat during occupied periods, non-residential passive solar heating systems must be carefully designed. Studies of passive solar cooling technologies indicate that ventilative, evaporative, and radiative cooling can meet most of the sensible building cooling load in many climates.

Daylighting is now being used successfully with electric lighting controls to significantly reduce both lighting and cooling energy use in non-residential buildings. Although there are fewer examples of the use of non-residential passive solar design techniques, it is clear that the cost-effective performance results and positive user-response to these existing buildings will increase industry acceptance and lead to increased application of these passive solar strategies.

Significant research and development in the areas of variable transmittance and insulating glazings will allow solar energy to supply more of the heating, cooling, and lighting energy requirements in an increasing variety of buildings and climates. The success of low-emittance glazing is a good example of a marketable product resulting from recent research efforts.

The ability to increase the thermal capacity of common building materials will offer additional opportunity for the efficient use of solar energy, especially in retrofit and non-residential applications.

The integration of these advanced materials with other building systems and the development of controls to regulate and distribute light and heat are necessary to achieve high performance passive solar buildings, especially in the non-residential sector.

The overall level of understanding has increased in the last decade, allowing successful passive solar applications with increased confidence even in marginal climates and larger, more complex buildings. The development of design tools and guidelines that have been accepted by the building industry has had a major influence on the application of these technologies and can assure the continued growth of passive solar energy as a viable option for building design.

VIII. BIBLIOGRAPHY

1. F.H. Morse, Office of Solar Heat Technologies, United States Department of Energy, "Overview of Solar Thermal Application in the United States". Presented at International Symposium on Thermal Applications of Solar Energy, April 8th, 1985.

2. J.N. Swisher and J.J. Duffy, "Measured Performance of 50 Passive Solar Residences in the United States", Proc. Passive and Hybrid Solar Energy Update, United States Department of Energy, Washington, D.C., USA, 1983.

3. R. Lutha, P.G. Rockwell, W.J. Fisher, and H.T. Gordon, "The DOE Passive Solar Commercial Buildings Programme: Preliminary Results of Performance Evaluation", Proc. Passive and Hybrid Solar Energy Update, United States Department of Energy, Washington, D.C., USA, 1983.

4. F.H. Morse, "Passive and Hybrid Solar Energy Programme Status, "Presentation to Joint US/UK Bilateral Workshop on Passive Solar Design", Newbury, U.K., 1984.

5. S.E. Selkowitz, A. Hunt, S.M. Lampert, and M. Rubin, "Advanced Optical and Thermal Technologies for Aperture Control", Proc. Passive and Hybrid Solar Energy Update, United States Department of Energy, Washington, D.C., USA, 1984.

6. D.K. Benson and C. Christensen, "Solid State Phase Change Materials for Thermal Energy Storage in Passive Solar Heated Buildings", Proc. Passive and Hybrid Solar Energy Update, United States Department of Energy, Washington, D.C., USA, 1983.

7. H.T. Gordon, J. Estoque, K. Hart, and M. Kantrowitz, "Nonresidential Buildings Programme Design and Performance Overview", Proc. Passive and Hybrid Solar Energy Update, Washington, D.C., USA, 1984.

8. "Comprehensive Review of Passive and Hybrid Solar Energy in the United States", United States Department of Energy, Washington, D.C., USA, 1984.

9. "Comprehensive Review of Solar Thermal Technology in the Unites States", United States Department of Energy, Washington, D.C., USA, 1984.

10. *Solar Architecture Solaire*, Proceedings of the International Solar Architecture Conference, held in Cannes, France, December 13th-16th, 1982.

D. SOLAR THERMAL TECHNOLOGY

I. INTRODUCTION

Archimedes, the third century B.C. Greek physicist may have been the first to chronicle the invention of solar concentrators, that is, use of reflective surfaces to create very high temperatures. Although known and experimented with for centuries, the earliest practical solar thermal systems using concentrating collectors were built in Europe and the United States in the late 1800s and early 1900s.

II.   TECHNOLOGY DESCRIPTION

Five different systems may be described as solar thermal technologies: parabolic troughs, parabolic dishes, central receivers, hemispheric bowls and solar ponds. In the last decade, most research and development has been devoted to the parabolic trough and dish and central receiver technologies, and troughs and dishes have advanced enough to be used in commercial projects supported by government investment incentives. Some development has occurred for solar ponds, artificial ponds that employ a heavy salt water concentration to collect and trap solar energy for industrial process heat and electricity. Solar ponds operate at much lower temperatures (85°C to 100°C) than the other solar thermal technologies, which focus radiation to increase its energy density. These systems convert solar radiation into thermal energy which can be used to produce shaft power for mechanical or electricity applications, or heat for commercial, industrial and agricultural needs.

High temperature solar thermal technology uses the principle of concentration to increase the work potential of the radiation collected. In such systems, solar collectors either reflect - or in some cases refract (bend through a transparent lens) - the direct rays of the sun on to a small target or receiver area). The concentrated solar radiation raises the temperature of the receiver from 150°C to 1700°C.

Figure 16 illustrates the basic concentrator configurations and shows photographs of three systems installed in the United States that employ these technologies.

The five basic components of high temperature solar thermal systems are the concentrator (collector), the receiver, the energy transport system, the thermal energy conversion system, and the control system. A sixth component in most systems is an energy storage sub-system. In solar ponds, the collector, receiver and storage systems are one.

Three basic types of solar thermal system have been examined extensively in the last decade:
—   Parabolic troughs;
—   Parabolic dishes;
—   Central receivers.

Parabolic troughs usually track the sun in one axis only and are referred to as *line focus systems.* The next two use two-axis tracking to follow the sun's movement and are referred to as *point focus systems.* Systems employing parabolic dishes or troughs are described as *distributed receiver systems* since each concentrator has its own attached receiver, hence a distribution

FIGURE 16
SOLAR THERMAL TECHNOLOGIES

CENTRAL RECEIVER ELECTRICITY PRODUCTION

 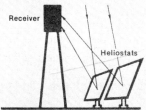

PARABOLIC DISH ELECTRICITY PRODUCTION

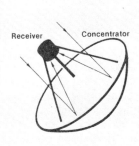 

PARABOLIC TROUGH PROCESS HEAT PRODUCTION

 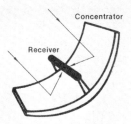

Source: United States Department of Energy.

of receivers exists throughout the field of collectors. Other systems that have received far less attention are the hemispheric bowl, solar ponds, and the line-focused fresnel lens.

Parabolic Trough

Parabolic troughs consist of a metal framework supporting a parabolic substrate covered with reflective material or a curved glass/metal or all-glass mirror. In some advanced systems the basic mirror modules are 2.4 meters wide and 6 meters long. Eight such modules can be linked together in a common tracking and drive unit. They are usually one-axis tracking collectors that concentrate approximately 40 suns (concentration ratio or number of suns refers to the area of the reflector compared to area of the receiver) on a receiver tube positioned along the focal line of the parabola. Although the linear focus concentration can also be achieved using a refractive lens, reflective systems dominate the technology. A piping system with a working fluid transports the collected energy from the receiver tubes to the point of use. The receiver tube is encased in glass and may be evacuated to increase performance. Either water or oil is used as the heat transfer medium. Due to lower operating temperatures, up to 400°C, troughs are usually considered most suitable for industrial process-heat applications. Nonetheless, they have been employed to power irrigation pumps or air-conditioning equipment for buildings and to make electricity for sale to utility companies.

Parabolic Dish

A parabolic dish module consists of a reflector and the structure necessary to support it, a tracking system on two axes, a receiver and energy converter at the focal point. In current technology metal/glass mirrors or aluminized plastic films are used for the reflectors. Most dishes have their concentrator mounted on a single pedestal. The energy at the receiver may be transported to supply heat to a thermal application, converted to electricity, or used for both in a total energy system. Two methods for generating electricity with a dish exist: 1) collecting heat from a field of dishes and piping it to a ground-mounted heat engine/generator set or 2) mounting the heat engine/generator at the focal point of the dish itself. A complete dish module may be used in a stand-alone configuration or grouped into a larger, multi-module system.

Dishes have similarities to central receivers and parabolic troughs. Like central receivers, two-axis tracking increases the energy collected and allows very high concentration ratios from one hundred to several thousand suns. The high temperature potential (1700°C) also permits use of Brayton cycles. Yet, like troughs, the basic dish module is small, allowing flexibility in meeting requirements of various applications, including industrial process heat.

Central Receiver

The central receiver is the largest scale solar thermal technology. In a central receiver system, the receiver is mounted atop a tower surrounded by or next to a large field of nearly flat tracking mirrors called heliostats. In current technology, receivers are arrays of metal tubes through which pass a heat transfer fluid. The heliostats that follow the sun and reflect and concentrate the solar radiation on to the receiver are computer-controlled. A large system, which in effect simulates a large parabola on flat ground, would employ thousands of heliostats, but only one receiver. The fluid in the receiver is piped to the base of the tower where it may be converted by a heat engine (steam turbine) to make electricity or may be used as heat for industrial purposes. Besides water, heat transfer fluids include molten salt or liquid sodium. Advanced systems may utilize air at very high temperatures with a Brayton cycle conversion system or micro particles of carbon for direct absorption. Central receiver systems can produce working fluid temperatures of 1500°C or higher. A thermal storage system is usually required to start up the system and smooth out performance during transient weather conditions. The ease of storing thermal energy from a central receiver system makes this technology ideally suited for utility applications where storage increases the plant capacity factor. It is expected that an economic module must exceed 10 MW$_e$ in scale, compared to tens of kilowatts for dish and trough technologies.

Solar Ponds

Solar ponds collect and store solar heat in large bodies of water. A typical configuration is the gradient pond where a salt density gradient below the surface, acting as an insulating barrier traps the incoming solar heat. Convection of the water warmed up at the bottom by the radiation absorbed there is inhibited by the barrier. This results in a thermal gradient and a lower surface temperature, thus reducing losses through evaporation, conduction, convection or radiation. Ponds can also be designed with other types of insulating barriers. Bottom temperatures can be as light as 100°C. The thermal energy is extracted with large heat-exchangers and can be used in a number of applications ranging from seasonal storage of heat for space heating or low temperature process heat applications, to electricity based on appropriate thermodynamic cycles. To be economically attractive, such systems may have to cover large areas of land.

III.   **APPLICATIONS**

Solar thermal technology has a high level of potential because the concentration-based systems provide thermal energy at elevated temperatures or are easily configured to produce electricity, thereby fitting well with conventional technologies. They offer further advantages due to quick construction times and modularity. Solar thermal technology can satisfy a diversity of energy needs because it:

—   has the potential to provide heat or make electricity at almost any scale (from kilowatts to hundreds of megawatts);

—   has established the highest efficiency for conversion of solar radiation to electricity (31.6%) and to heat (80%);

—   is compatible with conventional energy systems in existing plants;

— can provide very high-temperature heat without a chemical, mechanical, or electric-to-thermal conversion step;

— lends itself to energy storage systems that employ latent heat.

Through government and/or industry sponsored projects, the technical feasibility of high temperature solar thermal systems has been demonstrated for four key applications:

— industrial process heat;

— electricity production;

— irrigation pumping;

— desalination.

Research and development for industrial process heat and electricity systems have far exceeded that for other applications, including production of fuels and chemicals with solar flux or disposal of hazardous waste materials, which remain in exploratory phases.

Industrial Process Heat (IPH)

Solar thermal energy systems are capable of providing heat in a wide range of temperatures for many industrial applications. In the United States, approximately 50% of industrial processes require heat less than 300°C. Line-focus concentrator systems are ideally suited to provide for processes that require hot water, low pressure steam, or hot air in this lower temperature range. Lower temperature systems are applicable to the food processing, anodizing, and refining industries. Besides these industrial process heat applications, solar thermal technology may also be used for drying and dehydration of agricultural produce and to provide steam for thermally enhanced oil recovery. Central receiver and parabolic dish systems have been shown to produce lower temperature steam also, but these two-axis, point focus systems are expected to satisfy demand at temperatures above 300°C. Industries where these latter technologies have potential include primary metals, glass, cement, paper, container, chemical, brick and clay.

Electric Power

In many industrialized nations, as much as one-fourth of primary energy supply is used to generate electricity. Utility interest in new energy technologies to meet future supply requirements is responsible for a significant share of the research being conducted on solar thermal electric systems. Such systems have successfully demonstrated their feasibility as large-scale utility plants or small stand-alone units, and the first economic applications will be for peak power production and remote power systems. Solar thermal electric systems generally require the availability of high-grade maintenance staff. This limits their application for small-scale remote electricity production.

Central receivers, parabolic dishes and troughs have all been developed and tested for solar thermal electric applications. The most developed technology is the parabolic trough, which was intended to meet IPH needs. The largest installed solar electric power plants use trough technology which may well have the potential to be cost competitive without special economic incentives for certain near-term high value electric power applications. Their ultimate competitiveness will depend on the development of other solar electric technologies.

Central receiver and dish—the point-focus—technologies have excellent potential to satisfy demand for electricity. They can produce higher temperatures than trough systems thereby increasing operating efficiencies of heat-powered generators. Parabolic dish modules have been demonstrated with the engine/generator mounted on the ground in the collector field to use the thermal energy from a large number of dishes, and with an engine/generator mounted on each dish. Individual dish modules can be used to meet electric power needs in remote sites when the requirement for power ranges from 15 kW_e to 50 kW_e. Many modules can also be grouped for electric power applications requiring output in the range of 1 MW_e to 5 MW_e.

Like parabolic dishes with individual heat engine/generators, central receiver systems do not require extensive field piping and associated pumping power, nor do they suffer the same degree of thermal energy losses as parabolic trough or central engine dish systems. Central receiver systems have been shown to operate in temperature ranges that match those of conventional coal and nuclear power plants. Central receiver plants also interface easily with thermal storage sub-systems and track utility peaks well.

Irrigation Pumping

Irrigation pumping represents an important energy need in both developed and developing countries, and irrigation sites are often in rural areas that lack extensive utility grids. For these general reasons and those listed below, some of the first demonstrations of solar concentrator systems were for irrigation systems:

—   high levels of insolation coincide with seasonal demand for water;

—   irrigation systems usually require less than 300 kW;

—   irrigation can be scheduled to match solar availability;

—   thermal storage can be incorporated to satisfy more extended pumping demands;

—   non-irrigation demands (drying, heating, electricity) can be satisfied in off-season periods.

The most common system design uses parabolic trough collectors to provide thermal energy to an organic Rankine cycle (ORC) turbine. The shaft power can be used to operate either a pump or an electric generator.

Desalination

Providing fresh water supplies is an important demand for energy in many arid parts of the world with poor quality or insufficient water. Currently, most commercial-scale desalination plants are powered by fossil fuel power systems or by existing electric grids. Since most arid regions have excellent solar resources, a signficant potential exists for solar power plants to operate new desalination facilities, either for small stand-alone plants, or for large facilities providing water for towns and cities.

Aside from simple solar stills used successfully in many countries, commercial-scale solar desalination systems have only been demonstrated in a few locations. Overall system design for a solar powered unit may not differ significantly from a conventional desalination plant if the solar energy is converted to heat to evaporate water or make electricity. Research and development of solar desalination plants has concentrated on unconventional designs that maximize energy efficiency of the desalination system. The United States-Saudi Arabia SOLERAS programme constructed a demonstration in Yenbu, Saudi Arabia, using parabolic dishes to power an experimental freeze desalination technology. Other designs that came from this programme combined reverse osmosis with electrodialysis systems or used two-stage flash evaporation units.

IV.   STATUS AND ACCOMPLISHMENTS

Status in 1974

One hundred years ago European and American scientists and engineers developed operable solar parabolic dish and trough concentrator systems, but these systems failed commercially because low-cost oil and natural gas began their rise to energy dominance at that time. The last sizeable solar power system was built in 1913 in Meadi, Egypt. The only other application of the technology was in high temperature research facilities in France, Japan and the United States, that used solar furnaces. These facilities were built with a large reflector and small receiver to concentrate many "suns" and create very high temperatures. Two furnaces that existed in Japan were for fundamental materials research, while the others were originally developed for weapons effects testing. The Odeillo solar furnace in France is the largest of these, using 63 heliostats to focus energy into a parabolic concentrator mounted on the side of an eight story research facility. It concentrates 20 000 suns, producing temperatures up to 3 825°C. Besides construction of these advanced research facilities, little was accomplished in the solar thermal concentrator field until the 1970s.

Once interest in solar energy research was restimulated by events in 1974, a whole new field of scientific and engineering endeavor was initiated in the solar thermal area. Heat engine technology for solar systems had not been considered for many decades; therefore, researchers began their work without any significant data base. Only fundamental concepts were well understood and industrial capability to design and build systems did not exist. Reflectors were common glass mirrors, and there were no designs for receivers to collect and aide transport of thermal energy. No central receiver had ever been used to produce electricity, and only a few conceptual studies had been conducted on large systems.

Despite the poor level of knowledge that existed, in the early 1970s solar thermal technology appeared to have great potential, and significant RD&D programmes were begun in many countries. From a practically non-existent technical base, hundreds of systems have been installed in less than a decade, and highly successful test results obtained. Some technologies have advanced to the point where private companies have just begun to introduce them commercially.

Accomplishments

Several research, development and demonstration programmes carried out in IEA countries, frequently with international cooperation, have made important contributions to the advancement of solar thermal technologies. The list of key projects includes the IEA Small Solar Power System (SSPS) project at Almeria, Spain, the European Communities EURELIOS Project in Sicily, the French projects Themis and Pericles, the Sunshine Project in Japan, the Spanish CESA-1 project and the United States Depart. of Energy solar thermal programme. Table 8 provides data on key characteristics for a number of solar thermal energy installations around the world that have been constructed as experiments, demonstrations or commercial projects. The successes these systems represent occurred because of a high degree of sustained progress in a large number of scientific and engineering areas pertaining to solar thermal technology.

Parabolic Troughs. Parabolic troughs were the first of the solar thermal technologies to reach an advanced state of development. The SSPS project fielded two different parabolic trough technologies: one type tracked in a single axis; the second tracked in two axis. Lessons learned were: 1) that field thermal losses could be lowered 30% to 50% through better design of piping and insulation and better handling of thermal inertia; 2) that optical performance depends on how materials react to prolonged exposure; 3) that local meteorological and site conditions can have a significant impact on system performance; and 4) that well-designed and matched power conversion equipment is essential.

Numerous countries developed this technology, but manufacturers in the United States have constructed the most systems, advancing through four or five generations of collector technology and designing modular industrial process heat systems. Over 100 000 m^2 of trough collectors were installed in experiments and demonstrations in the United States alone. Tracking controllers have been perfected, fluid handling improved drastically from the earliest systems, and the manufacturing of suitable glass reflectors in the parabolic shape perfected. In addition, manufacturers have designed the latest generations of technology for mass production.

TABLE 8
CHARACTERISTICS OF SOLAR THERMAL PROJECTS
Characteristics of Operational Solar Thermal Industrial Process Heat Plants

Plant	Owner	Location	Lati-tude (deg.)	Application	Load Temp. (°C)	Collector	Array Area (m²)	Placed in Operation	Monitored period
Caterpillar Tractor	Caterpillar Tractor Co.	San Leandro, CA, USA	38	Preheat water for washing parts	91	Parabolic trough	4 684	1982	19 months
Dow Chemical	Dow Chemical Co.	Dalton, GA, USA	35	Process steam— unfired boiler	187	Parabolic trough	929	1981	36 months
Home Laundry	Home Cleaning and Laundry Co.	Pasadena, CA, USA	34	Hot water (71°C) or 100-110 psig steam in commercial laundry	166	Parabolic trough	604	1982	18 months
Ingelstad	Vaxjo Local Authority	Vaxjo, Sweden	57	Space heating and domestic hot water for 52 single family houses	71	Parabolic trough	1 390	1979	48 months
Lone Star Brewery	Lone Star Brewery	San Antonio, TX, USA	30	Process steam for brewery—unfired boiler	177	Parabolic trough	878	1982	12 months
Southern Union IPH	Southern Union Gas Co.	Lovington, NM, USA	33	Process steam for petroleum	191	Parabolic trough	937	1982	13 months
USS Chemicals	United States Steel Chemicals	Haverhill, OH, USA	41	Process steam for chemical plant — unfired boiler	151	Parabolic trough	4 680	1983	8 months

TABLE 8 (Continued)
CHARACTERISTICS OF SOLAR THERMAL PROJECTS
Characteristics of Operational Solar Thermal Industrial Process Heat Plants

Plant	Owner	Location	Latitude (deg.)	Application	Load Temp. (°C)	Collector	Array Area (m²)	Placed in Operation	Monitored period
ARCO EOR	Atlantic Richfield Co.	Bakersfield, CA, USA	36	Steam for enhanced oil recovery	286	Central Receiver	1 620	1983	18 months
Dyeing factory	NEDO	Ichinomiya, Aichi, Japan	35	Cascaded heat demonstration process for dyeing		Parabolic trough	532	1984	24 months
CESA-I	Ministry of Industry and Energy, Spain	Almeria, Spain	37	Experimental Solar Power Plant	510	Central Receiver	17 000	1983	36 months
SSPS-DCS1	International Energy Agency	Almeria, Spain	37	Experimental steam power plant	280	Parabolic trough (ACUREX)	2 674	1981	36 months
SSPS-DCS2	International Energy Agency	Almeria, Spain	37	Experimental steam power plant	280	Parabolic trough (MAN)	2 688	1981	36 months
SSPS-DCS3	International Energy Agency	Almeria, Spain	37	Experimental steam power plant	280	Parabolic trough (MAN)	2 240	1984	36 months
14 Dish-Power- System	New South Wales Government	White Cliffs, Australia	-32	Small community co-generation steam power plant	550	Parabolic dish	270	1982	18 months

TABLE 8 (Continued)
CHARACTERISTICS OF SOLAR THERMAL PROJECTS
Characteristics of Operational Solar Thermal Industrial Process Heat Plants

Plant	Owner	Location	Lati-tude (deg.)	Application	Load Temp. (°C)	Collector	Array Area (m²)	Placed in Operation	Monitored period
S.T.E.P.	Georgia Power Co.	Shenandoah, GA, USA	33	Co-generation of electrical power and process steam and cooling - 2571 absorption chiller	440	Parabolic dish	4 400	1982	18 months
SOLAR ONE	Southern Calif. Edison Company	Barstow, CA, USA	35	Experimental steam power plant	510	Central Receiver	72 538	1982	26 months
SEGS 1	Luz Solar Partners	Barstow, CA, USA	35	First of a kind electric	300	Parabolic trough	100 000	1984	12 months
SEGS 2	Luz Solar Partners	Barstow, CA, USA	35	Commercial electric	300	Parabolic trough	165 000	1985	1 month
Vanguard Advanco	US DOE	Palm Springs, CA, USA	34	25 kW$_e$ prototype	700	Parabolic dish	92	1983	24 months
Solar Part 1	La Jet Energy Co.	Warner Springs, CA, USA	33	4 MW$_e$ first of a kind	400	Parabolic dish	30 300	1984	24 months
Dish 1	Southern Calif. Edison Company	Barstow, CA, USA	35	25 kW$_e$ prototype	700	Parabolic dish	92	1985	3 months

Through these advances, system performance has steadily improved. Some of the first trough systems were providing annual solar radiation conversion efficiences of only 6%. This figure improved to 27% in the better demonstration projects installed in the early 1980s. The last few years of development have improved parabolic trough collectors so that thermal efficiencies that ranged from 17% to 43% in field tests would now be in the range of 65% to 70% with fluid output temperatures at more than 250°C. These gains are the result of better design of fluid-transport systems, application of anti-reflective coatings on absorbers (8% to 18% gain in collector efficiency) and perfection of components like rotary joints that can affect system availability and reliability. Latest generation systems have shown availabilities exceeding 95%.

Parabolic Dishes. Research and development of parabolic dish technology has led to significant improvements in many technical areas. Regarding concentrator design, technology has progressed on parallel tracks: one using glass-metal technology for the reflectors and the second using metalized (aluminized and silver) plastic film technology. Early glass-metal concentrators cost US $1 300/m². Cost projections for current models (11 meter diameter) are now in the US $300/m² range. The projected cost of aluminum polyfilm concentrators has dropped from approximately US $1 000/m² to below US $200/m². Weight per unit area of reflector has also been reduced significantly.

Power conversion for dish electric systems has also progressed on two fronts:

1) conversion of the thermal energy to electricity at the focal point of each module; and 2) transport of the thermal energy to a central point for conversion. For the second approach, several large multi-dish systems have been installed in Australia, the United States and the Middle East. Experience with dish-mounted engine/generators is more limited but indicates highly promising performance. The most extensively-tested unit was constructed in the United States under Department of Energy sponsorship. By April 1985, the Vanguard Parabolic Dish Stirling Module (25 kW$_e$ in Palm Springs, California, United States) had logged 1 200 hours

TABLE 9

VANGUARD DISH STIRLING MODULE PERFORMANCE

Maximum Output Capacity	25 kW$_e$
Gross Conversion Efficiency	31.6%
Net Conversion Efficiency	29.4%
Net Daily Average Efficiency	25.23%
Gross Daily Averge Efficiency	27.4%
Net kWh/m²/day	2.52
Gross kWh/m²/day	2.74
Net Output of a Parabolic Dish/day	238 kWh
Gross Output of a Parabolic Dish/day	258 kWh

of testing with its power flowing to a utility grid. The eight technical records presented in Table 9 have set a standard for future technology development. The builder of this system estimates that the unit would establish an average annual conversion efficiency of 23% with a capacity factor of 25.68%.

The largest dish system with a central energy transport design that has been extensively monitored is also in the United States. Since 1982, Georgia Power Company has operated a multi-use (total) energy system with 114 dishes and a capacity of 3 MW$_{th}$. The system output is converted to provide 400 kW$_e$ of electricity, 250 tons of air conditioning and 1 340 pounds of process steam per hour at 182°C. In steady state operation, efficiency of the solar collector field has reached 52.6%. (See Table 8, Figure 17).

No data were available on the performance of a multi-dish installation in Kuwait installed by a German company. Fifty-two dishes that rotate about a fixed receiver were installed in 1981. Each is five meters in diameter. Nor were data available on the performance of the largest dish installation built to date, a private project in Warner Springs, California, United States, named Solar Plant I and rated at 4.92 MW$_e$ (gross) using the thermal collection concept.

FIGURE 17
SHENANDOAH SOLAR COLLECTOR FIELD OPERATING EFFICIENCY

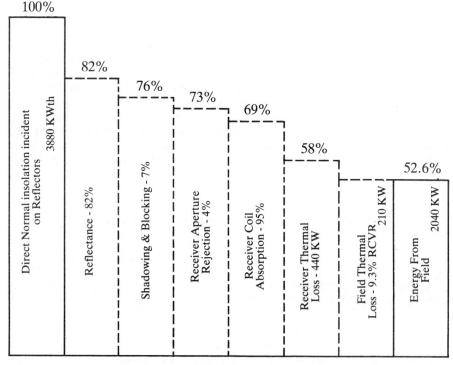

Source: Distributed Solar Thermal Annual Review, Williamsburg, Va, 1984.

Central Receivers: Industry, utilities and government research institutions in several leading countries have achieved a solid understanding of central receiver system concepts using different heat collection and transfer media: water/steam, molten nitrate salt, and liquid sodium. Through construction and operation of pilot-scale plants and key component experiments in France, Italy, Spain, and the United States, valuable experience has been gained in construction, system integration, start-up, controls, non-solar material and sub-system performance, and automated operation.

The SSPS central receiver system in Spain employed a cavity-type receiver with liquid sodium as the heat transfer medium to demonstrate the feasibility of using a sodium heat-transport system. The SSPS project also tested two receiver configurations (cavity and high flux external), and recorded the highest flux (solar radiation) concentration (8 MW/m^2) on a receiver.

Another significant advance was the construction in 1982 of the THEMIS central receiver system in France at Targasonne. Rated at 2 MW$_e$, it used 201 heliostats to concentrate solar radiation into a cavity type receiver. Molten salt was used as the heat transfer media *and* for thermal storage, giving the plant the capability of producing electricity for six hours without the sun. THEMIS, which completed its operation in June 1986, has provided excellent data on the largest molten salt system to have operated to date.

In 1979, the Spanish Ministry of Industry and Energy sponsored planning and construction of the *Central Electrica de Almeria* Project (CESA-1). Sandia National Laboratories provided technical assistance from the United States. CESA-1 is a 12 MW$_{th}$ pilot plant, using 300 heliostats built by two different Spanish manufacturers. The receiver transfers its energy with a water/steam loop to a molten salt storage system. Located next to the SSPS project in Spain, the plant began operating in 1983.

After initial tests on its water/steam receiver, the plant was refitted under the auspices of a Spanish-German programme (GAST) to test a metal air-type receiver. The original receiver has been reinstalled for performance tests to be conducted in 1986. Later, GAST will use it to test a new ceramic air-type receiver.

Under the Sunshine Project in Japan, two different pilot plants were built at Nio. The central receiver plant used a field of 807 small heliostats and a cavity-type receiver. Saturated steam powered the turbine generator, while pressurized water provided thermal energy storage for up to three hours operation. The second plant incorporated a unique design, combining heliostats with parabolic trough-type reflective receivers. When tests were concluded in March 1984, these two plants had operated for two and a half years, producing 1 235 MWh in 2 125 operating hours.

The United States has built the world's largest central receiver plant, a pilot-scale unit commissioned in Southern California in 1982. Rated at 10 MW$_e$ peak power, the receiver for Solar One is 13.7 meters in height, mounted on a 91.4 meter tower. The collector field, which surrounds the tower, totals 1 818 heliostats, each 39.9 m^2. In 1984, the United States Department of Energy team turned Solar One over to the sponsoring utility for operation as a utility power plant. The plant has exceeded its peak design output by 20%, successfully operated at night from storage, and demonstrated a capability of achieving an overall annual plant system efficiency of 13.0%. Table 10 summarizes key Solar One achievements.

TABLE 10
SOLAR ONE OPERATIONAL MILESTONES THROUGH DECEMBER 1985

Maximum Instantaneous Power Output	10.8	MW$_e$(net)
Maximum Energy Production Per Day	62.62	MWh(net)
Maximum Energy Generation Per Week	455.80	MWh(net)
Maximum Energy Generation Per Month	1 755.84	MWh(net)
Maximum Energy Generation Per Year	8 803.14	MWh(net)
Longest Operating Time in a Day	11.68	Hours
Longest Operating Time in a Week	73.02	Hours
Longest Operating Time in a Year	1 801.6	Hours
Largest Number of Heliostats in Service	1 818	= 100%

Source: Southern California Edison. Presentation by C. Lopez.
Diver, Richard B. Ed. *Proceedings of the Solar Thermal Technology Conference, Albuquerque, NM, June 17th-19th, 1986. Sandia National Laboratory, June 1986, p. 38.*

Research and development in the United States and Europe have shown that liquid sodium or molten salt systems could raise the efficiency of central receiver plants to 22%. This potential level of performance and the capability of storing energy for hours has prompted the development of a small number of commercial versions of a central receiver system. Their analysts have predicted that central receiver plants must be near 100 megawatts in size to produce electricity economically. Such a plant would require 6 000 heliostats 100 m^2 in area, a receiver 23 meters in height and a 136 meter tower. Land requirements would equal 3.1 million m^2. New studies are underway to analyse data gathered in the last five years and to re-evaluate commercial design options. Meanwhile, testing continues on key sub-system components such as receivers and automated control systems.

In the pilot central receiver plants built to date, heliostats have accounted for 40% to 50% of overall costs; consequently, lowering heliostat cost has been given high priority. Significant success has already been achieved, demonstrated by manufacturer quotations of US $200/m^2 for glass-metal heliostats compared to the installed cost of nearly US $900/m^2 in the 1970s. Heliostat performance has also shown rapid gains. With 1970s technology, the heliostat field at Solar One is averaging 95% to 97% availability and has periodically delivered 100% availability. Monitoring has shown that soiling of heliostat surfaces has an important effect on optical performance of the collector field, and new advances in development of heliostat washing techniques have also been forthcoming.

Solar Ponds: Solar pond technology is commercially-advanced in Israel but not in IEA Member countries. Countries such as Australia, Italy, Japan, Spain, Portugal, and the United States of America have on-going pond research. Portugal has an R&D programme in this area and operated a 1 000 square meter salt gradient pond (3.5 meters deep) for several years to heat agricultural buildings. Australia has developed an electricity-producing pond, 2 000 m² in size, which has generated power for more than 1 700 hours. The United Kingdom has researched solar ponds, and Japan is investigating them. Sandia National Laboratory in the United States has operated several test facilities for a number of years. The research and experimental work have shown that the salt gradients can be maintained and reliable power produced from these systems. Salt pond efficiency for thermal energy has been demonstrated to be about 15%. Conversion to electricity has been shown to be practically achievable; however, the overall conversion efficiency is about 1% to 2%. Research shows this conversion rate to be acceptable should pond costs fall as predicted.

Many countries have been strong participants in the technical progress and projects described above. Besides those mentioned, Germany and Switzerland have been key participants. As a result of a decade of the work in all the countries involved, solar thermal technologies have progressed to advanced development stages, well beyond proof of concept. For each of the main technological approaches described, experimental and demonstration installations have proved technical feasibility. While existing systems are further refined and improved to reduce high costs and increase performance and lifetimes, still newer systems and concepts are emerging from various research programmes.

Commercial Status

Solar thermal technologies are in varied stages of commercial development. Many concepts have been tested and rejected, others proven and dramatically improved. Some of these technologies are still at the advanced prototype stage where feasibility has been proved and the lessons learned ingested but more advanced testing of a new generation is necessary. Still others are ready to be placed in the field in a commercial-scale configuration. A few companies have advanced some solar thermal systems to initial commercialization stages, having successfully sold one or two large installations, or placed a nearly ready unit with potential customers for testing under normal operating conditions in the United States.

Until 1984, government research and demonstration projects provided the major market for high temperature solar thermal hardware for both electric and industrial process heat systems. The United States Department of Energy industrial process heat and irrigation demonstrations and the IEA

SSPS project absorbed nearly 100 000 m² of parabolic troughs. Industry has also produced approximately 120 000 m² of heliostats for the central receiver test and pilot plant facilities.

Sales of commercial energy systems using solar thermal concentrators have been limited to a few major projects, almost uniquely in the United States, in Southern California. Several favourable conditions exist in this part of the United States such as: peak electricity load caused by air conditioning coinciding with the maximum solar power, ample available desert space for solar plants, favourable climate conditions and the population's preference for renewable energy. One United States manufacturer successfully sold two private industrial process heat systems using parabolic troughs, one 1 900 m² and the other 5 600 m². After 1983, however, United States firms involved with trough systems ceased aggressively marketing their products. Falling oil and gas energy prices, the expiration of federal tax credits and the difficulty of selling systems with a payback longer that two or three years effectively blocked the market. While trough technology is ready for commercial introduction, it would benefit from additional R&D to improve collectors, explore low-cost production techniques and develop automated controls.

Parabolic trough systems have continued to advance despite the lack of development of an industrial process heat market. An Israeli firm with a California subsidiary selected trough technology for a venture into the third-party financed, independent energy-producer market in California. As a result, parabolic troughs are used in the two largest solar electric power plants in existence, SEGS I and II.

The plants are augmented by natural gas (a hybrid design) in order to maximize income from electricity sales to a utility under long-term, negotiated power-purchase contracts. SEGS I has a gross output of 14.8 MW_e from a collector field 70 000 m² in area; SEGS II is 30 MW_e (gross) with 160 000 m² of collector. These plants were constructed in 1984 and 1985. In 1986, the company expects to close financing of SEGS III, another 30 MW_e plant, and has plans to build several more.

Parabolic dish technology has been sold primarily for government demonstration or test programmes. No significant prospects for dish thermal sales appear close at hand. For dish electric systems, some early market penetration steps are being taken by firms in the United States. One large privately-financed dish electric project has already been installed, financed as an independent energy project to sell power to a utility in California. Seven hundred dishes 7.4 m in diameter (each 43 m²) provide thermal energy to a centrally-located steam rankine turbine/generator rated at 4.9 MW_e. The concentrator uses stressed polymer film for the reflector. Another United States company has built eight Dish-Stirling modules using a Swedish-developed 25 kW_e engine tested in Department of Energy projects. Constructed with metal/glass reflectors, four of these systems have been located with utilities for testing, while others are expected to be sold overseas. A 4.92 MW_e dish system privately installed in California has set a benchmark for dish systems with a centrally-located generator. The United States Department of Energy reports that the concentrators cost $15/m^2$ and the total system cost was $3 400/kW_e$. Company officials have observed that learning curve experience offers the potential to reduce costs of this system by as much as 40%.

Central receiver technology has successfully advanced through key experiments and analysis in five countries. Assessment studies are underway to evaluate past performance and reconfigure the technology for the next generation plant which would be a commercially designed system. Utilities in the southwest United States are prepared to begin a new series of design studies with Department of Energy funding in anticipation that this option could be ready for commercial deployment in the late 1990s. In 1986, several Western European companies formed a consortium to conduct a feasibility study for a new central receiver plant that could be marketed in a developing country with high energy costs and insolation levels. Technological developments in 1985 and 1986 in the areas of reflective surfaces and low-cost heliostat design have shown that a further 50% cost reduction for heliostats could be realized in one or two years.

Solar ponds appear to be advancing technologically and could be approaching commercial status in countries where active public and private programmes exist. Despite the fact that their large storage capacity compensates for their low conversion efficiency compared to other solar thermal technologies, the pace and scope of development is limited. Israel and the United States have completed the most testing. More data on performance is needed to evaluate the commercial prospects and estimate life cycle energy costs. Design studies in the United States have recently been completed for a 12 MW pond in southern California, the first phase of a 48 MW project that will sell power to a utility company.

An important indicator of the potential competitiveness of central receiver systems are heliostat costs because they accounted for as much as 40% of the system costs for systems designed in the early 1980s. In 1986, the United States Department of Energy will begin testing two prototype stressed membrane heliostats estimated to cost US $60 to US $80/m^2$ in mass production.

V.   ENVIRONMENTAL CONSIDERATIONS

Potential environmental, health, and safety impacts associated with the production and use of solar thermal energy systems appear in most cases to be negligible, easily mitigated, or site-specific. Pollutants arising during routine operation are relatively minor and can be controlled by methods that are used routinely in industry. Impacts on land-use, local ecology, and water use during construction and operation of a solar thermal system may be more significant, although still less than that for most conventional power generation technologies.

Large expanses of land are needed for solar thermal systems - approximately 2.6 km² to capture 100 megawatts of electricity. In addition to the problem of finding and purchasing such large tracts of land, the construction of any large facility such as a power plant will impact the local ecosystem. In truth, however, in dry sunny climates where this technology will be applied first, land availability is not expected to be a problem. The operation of a solar thermal power system is not likely to cause any pollution, with the possible exception of an accidental release or spillage of some working fluids that may be used for heat transport or storage.

VI. TECHNOLOGY OUTLOOK

Technological Prospects

The costs of solar thermal energy systems have declined dramatically in this last decade of development and deployment. Further technology development is predicted to lower costs by 50% or more, which could make the costs of solar thermal power attractive for deployment in the high value electricity market in the 1990s.

Although the pace of solar thermal R&D has slowed in recent years, government institutions and industry in some countries continue to make progress in key areas. The R&D underway is being devoted to development of full-scale systems and components and automation of production as well as generic research on materials, heat transfer, engines, reflective and absorbative surfaces and structures. Advances in any of these areas can frequently benefit any of the solar thermal technologies that use concentrators.

Annual efficiency of solar thermal technology has increased steadily, and state-of-the-art systems have reached thermal efficiencies of 80% under peak conditions. Earlier poor performance and outright system failures experienced have been caused by designs too simple or too complex to operate well. Shortfalls have also been due to higher than predicted thermal losses, poor resource predictions or changes in insolation and weather patterns, faulty installation of key components such as pipe insulation, and unexpected requirements for trace heating of thermal piping. Failures of solar components have involved control systems, mirrors, mirror coatings, absorbers and their selective coatings, glazings, valves and flexible hoses. A high percentage of solar system failures have been due to inadequacy of conventional, non-solar components. The technological prospects for continued progress in solar thermal technology will require that research be applied to well-designed systems and components. These failings are attributable to correctable causes, and their rate of occurrence appears to be decreasing. Consequently, the outlook for continued gains in performance appears favourable, and further improvements are expected.

Each new generation of parabolic trough technology has increased in performance and state-of-the-art system efficiencies now equal 65%-70%. Gains in reliability and performance have come from better designed and manufactured equipment. Design gains have been due to lower temperature applications, reconfigured field piping systems and better load matching. With availabilities increased to over 95% and control strategies refined, newer installations have consistently outperformed earlier ones. Furthermore, installation of very large field arrays in the last two years has led to further technology improvements (e.g. enlargement of trough apertures by 100%) and indicates that important potentials from manufacturing economies are realizable for solar thermal systems. Development of higher temperature heat transfer oils and less expensive reflectors should help parabolic trough technology advance further; however, its outlook is more dependent on manufacturing economies and the price of conventional energy.

Developments in parabolic dish technology have stimulated greater interest in this technology. The improvements have come from innovative concentrator designs that apply thin glass-metal and stressed membrane reflectors. New support and tracking structures are further reducing weight and cost. Heat engine research and development appears to be progressing and should satisfy the need for one or two more generations of this key sub-system. Private sector development of a dish/electric module has nearly equalled performance of state-of-the-art experimental systems. Advances in concentrators would also benefit systems for thermal-only applications: however, the more favorable technological outlook in the near-term appears to be related to dish-electric systems. A benchmark for systems with a dish-mounted engine/generator is a commercial prototype unit developed in the United States. Company representatives estimate current installed costs for their Dish-Stirling 25 kW_e module to be US $4 000/$kW_e$. When produced in series, these modules are expected to cost US $2 000/$kW_e$.

Over the last five years, excellent field data have been generated for central receiver systems and design teams in Europe and the United States are evaluating the results in preparation of the next generation of technology. Advances in receiver design, collector field design and operation, thermal systems, control strategies and automated operation have sustained industry, utility and government interest in this technology. Development of a new generation of heliostats based on stressed membranes and the potential of new techniques for applying silver to steel substrates promise much lower cost collectors for central receiver systems. Advances in receiver concepts appear imminent in Europe and the United States. The technical prospects appear to favor development of a next generation of technology with commercial attributes and annual efficiency above 15%.

Overall efficiency for the three principal solar thermal technologies must still be improved to increase their competitive chances. Improvements can be anticipated as a result of better components; however, the major research and development efforts will be on improving performance and reliability, manufacturing cost reductions, and lower labour requirements to operate the systems. A major thrust will be reduction of materials intensiveness of collector technology through innovative designs that provide stability and stiffness in light-weight structures using stressed membranes, tubular space frames, and new materials. Durable, high-performance, low-cost materials such as silvered polymer films and silvered steel foil are expected to replace glass in reflectors. These developments could reduce cost of heliostats and other reflectors to US $50/m^2.

Regarding conversion of radiation to heat, flux concentration in central receivers will be increased to reduce receiver size and heat losses. Eventually direct absorption receivers may appear in which concentrated solar energy is absorbed by the working fluid without tubular containment. Increased efficiency for systems with high concentration rations will depend on materials advances since current technology already exceeds the capability of existing materials. Plant designs will become more optimized and increasingly sophisticated and should further minimize thermal losses and inertia of the components. Sophisticated control systems will allow low-cost, reliable, unattended operation of such plants, a requirement widely recognized as essential to the future of these technologies. In the early stages of market penetration, designers will have to optimize the application of solar resources with conventional energy systems to satisfy total loads in the most efficient manner.

Longer-range research is aimed at developing new insights and opportunities through improved understanding of materials degradation processes, interface phenomena, analytical methods and tools for materials research, component and systems design and photochemical and thermochemical reactions. The longer-term outlook for the technology may be tied to its unique ability to combine high temperature solar energy and storage processes. The result should be development of new processes that take advantage of concentrated sunlight to drive thermochemical or photochemical reactions. Production of fuels and chemicals is especially attractive and should receive an increasing level of research attention given future needs for easily transportable, high-density fuels. Should these new approaches be perfected and become commercially available, the market penetration prospects for central receiver plants would improve significantly.

Outlook

Although major advances have been made to date, comparisons of the current capabilities of solar thermal technologies with long-term system and component goals show that progress will be required on all fronts to reach those goals. The required quantitative advancements in capital cost and annual efficiency are only one aspect of the improvements needed. More confidence is needed that these technologies can perform reliably with high system availability and low maintenance costs. Given the solid technology base already established and on-going efforts, there are many reasons to expect confidence to grow.

The larger-scale solar thermal technologies are likely to have difficulty achieving commercial readiness in the current economic environment. For example, European and American companies feel that central receiver technology is ready for a new system demonstration in the 30 to 100 megawatt size range. The cost of such a project would begin at US $90 million or more which represents a risk that industry is not prepared to undertake alone. Distributed solar thermal technologies may reach commercial status more easily since they are smaller in scale. However, since individual units may not provide all the experience needed for full commercialization to proceed, some of these systems may also require larger, more expensive multi-unit installations to demonstrate commercial readiness. It is not anticipated that government projects will account for large purchases in the near future and future sales in any volume should come from private commercial projects. Should new approaches for reflective surfaces and low-cost heliostats be perfected and become commercially available, the market penetration prospects for central receiver plants would improve significantly.

That few specific plans exist for new demonstration or commercialization projects for solar thermal electricity in IEA countries reveals the degree to which the pace of development of these technologies has slowed in the last four years. Although industry involvement has declined proportionally, those companies still engaged in this field continue to believe in the future economic viability of their technology. No significant cost or technical barriers appear to be insurmountable. For trough technology, which is closest to commercial status, the economics of conventional energy appear to be the only major barrier.

In the United States, reduced R&D funding has placed central receiver development on a time-table that would not allow it to reach a state of readiness for commercialization until the mid-to-late 1990s. Dish-electric technology in small modules (25 kW$_e$ to 50 kW$_e$) appears in a position to advance to a competitive status in the 1990s. Interest in this technology for larger-scale systems is also evident. Thermal applications may follow, but most of the interest appears keyed to electricity output at this time.

If stated R&D goals regarding performance and cost are approached or achieved, and if reliability and durability of system components continue to be improved, solar thermal technologies would become a viable source of alternative energy and would begin entering high value energy markets in the mid-to-late 1990s.

VII. FINDINGS

Major technological advances have been made in solar thermal technologies since 1974, primarily for industrial process heat and electricity generation applications, but also for desalination and irrigation systems.

Large improvements in thermal performance are still possible for each of the solar thermal technologies. Industry expects a strong government role to remain essential for development of these technologies because of falling energy prices and high technical risks in certain areas.

A need still exists for analysis and design tools in order to improve systems performance prediction and design. In addition, data remain insufficient for reliable prediction of system lifetimes.

Potential markets for solar thermal technologies appear vast, especially for high temperature direct process heat. Long-term markets should develop that take advantage of unique characteristics of concentrated solar thermal flux.

Due to favorable tax and investment incentives, three large (more than 5 MW$_e$) solar thermal electricity power plants have been constructed with private finance in the United States. In other IEA countries with experience in operation of such systems, no specific demonstration or commercial plans appear to exist.

Solar thermal electric systems are not presently competitive with electricity generation by conventional utilities, but in some remote applications they could be cost-competitive with high-cost conventional systems. A high degree of disparity exists with regard to the technological readiness of each of the candidate technologies, and more research and development will be required to make these technologies fully competitive in the future. At present, such plants are being designed to meet peak load requirements of electric utilities where they can displace high cost conventional fuels. Solar thermal electric systems lend themselves to such applications because their power output coincides well with daily and seasonal peak load conditions of utilities in hot sunny climates.

VIII. BIBLIOGRAPHY

1.   Create New Energy, New Energy Development Organisation, Tokyo, Japan, September 1984.

2.   Kyodo Press, Japan 8, November 1984.

3.   Denki Shimbun, Japan, 2 November 1984.

4.   Brown, Kenneth C., *Handbook of Energy Technology and Economics*, John Wiley and Sons, New York, 1984, p. 618.

5. International Energy Agency, Small Solar Power Systems Project (SSPS), *The IEA/SSPS Solar Thermal Power Plants, Volume 1: Central Receiver System (CRS)*, Springer Verlag, Berlin, Heidelberg, New York, Tokyo 1986.

6. International Energy Agency, Small Solar Power Systems Project (SSPS), *The IEA/SSPS Solar Thermal Power Plants Volume 2: Distributed Collector System (CRS)*, Springer Verlag, Berlin, Heidelberg, New York, Tokyo, 1986.

7. International Energy Agency, Small Solar Power Systems Project (SSPS), *The IEA/SSPS Solar Thermal Power Plants Volume 3: Site Specifics*, Springer Verlag, Berlin, Heidelberg, New York, Tokyo, 1986.

8. International Energy Agency, Small Solar Power Systems Project (SSPS), *The IEA/SSPS Solar Thermal Power Plants, Volume 4: Book of Summaries*, Springer Verlag, Berlin, Heidelberg, New York, Tokyo, 1986.

9. Solar Energy Industries Association, *Energy Innovation: Development and Status of the Renewable Energy Industries*, Washington, D.C., 1985.

10. LaPorta, Carlo, Erikson, Jeffry and Seielstad, Harold, Solar Energy Industries Association, *A Review of the Solar Thermal Power Industry: Future Outlook*, Washington, D.C., 1986.

11. United States Department of Energy (DOE/CE-T13), *Solar Thermal Technology: Annual Evaluation Report—Fiscal Year 1984*, Washington, D.C., 1985.

12. Solar Energy Research Institute, United States Department of Energy (SERI/SP-271-2511), *Solar Thermal Technical Information Guide*, SERI, 1985.

13. Solar Energy Research Institute, United States Department of Energy, *Design and Performance of Large Solar Thermal Collector Arrays: Proceedings of the International Energy Agency Workshop*, San Diego, CA, June 10-14, 1984, SERI, 1985.

14. Deutsche Forschungs- und Versuchsanstalt für Luft- und Raumfahrt e.V. (DFVLR), SSPS - Results of Test and Operation (SSPS-SR7), Köln, 1985.

E. PHOTOVOLTAIC TECHNOLOGIES

I. INTRODUCTION

The "photovoltaic effect", which causes electricity to be produced when light strikes certain materials, was first discovered by French physicist Edmund Becquerel in 1839. It was Albert Einstein, however, who found that photons, or tiny particles of light, can interact with the electron shell surrounding an atom's nucleus to cause a free stream of electrons, that is, an electric current. The first photovoltaic cells were little more than a scientific curiosity until Bell Laboratory scientists took a major step in 1954 toward turning the concept into a practical power source when they succeeded in producing a solar cell from pure silicon which had an efficiency of 4%.

Energy from the first photovoltaic systems was about one hundred times more expensive than that generated by conventional energy sources. Today, this energy is only five to ten times as expensive, and costs continue to decrease. The efficiencies of the cells have steadily improved, and new materials and fabrication processes are being developed which are believed to have great promise with respect to cost and performance.

A major industry has emerged around photovoltaic technology whose growth is evident from the steady increase in module production—from 1 MW_p in 1978 to 24 MW_p in 1985. While photovoltaics was first sustained by the space industry, terrestrial applications have long since assumed greater importance. The major applications are consumer products, and stand-alone utility systems, navigational aids, water pumping, and communication systems.

II. TECHNOLOGY DESCRIPTION

Photovoltaic systems convert solar radiation to direct-current electricity without moving parts or thermal energy sources. The basic element is the photovoltaic or solar cell which is comprised of semi-conductor materials which have both negative and positive (the majority) charge carriers (called n and p layers respectively). When photons from ultraviolet, visible, and near-infrared light enter the cell, electrons in the semi-conductor material are freed, and an electric current is generated (See Figure 18).

A typical cell is about 100 cm^2 in size and can produce one peak watt of electrical energy. A peak watt (W_p) is the power generated on a clear day when the sun's rays strike perpendicular to the cell. Groups of cells are mounted on a rigid plate and wired together to form a module, typically 1 m^2 or less in size with a generating capacity of 50-100 W_p. Commercial solar modules convert between 9% and 12% of the sunlight that strikes them

FIGURE 18

DIAGRAM OF A BASIC P-N JUNCTION PHOTOVOLTAIC CELL

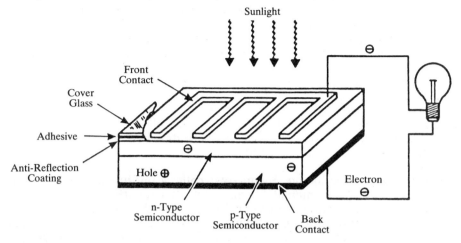

Source: *Photovoltaics: Technical Information Guide.* Solar Energy Research Institute, February 1985, p. 2.

into electricity (90-120 watts/m²), while efficiencies of 11% to 22% have been achieved in the laboratory. Groups of modules are electrically and physically connected together in an array.

The other system components, called balance-of-system (BOS) components, may include (depending on the application) power conditioners to convert direct current to alternating current, current and voltage regulators, batteries for energy storage, power controls, and supporting structural parts. BOS components, together with indirect costs (design, construction management, interest during construction), can amount to over 50% of the system cost.

Collector modules are generally assembled in a flat-plate configuration. A special approach utilizes concentrating collector modules (described later in this section) which are mounted on a sun-tracking mechanism. Flat-plate PV modules can also be mounted on tracking systems. Concentrator cells, since less area is required, can utilize techniques to improve efficiency that would otherwise prove too costly. Their output is further enhanced because the cells are directed at the sun the entire day.

Solar cells can be made from a number of materials and fabricated in a variety of designs. Some of the most prevalent semi-conductor materials employed for PV cells are single-crystal silicon, amorphous silicon, polycrystalline silicon, copper indium diselenide, and gallium arsenide. Single crystal silicon has been the semi-conductor material most often used in PV cell production since 1980. However, its share of the market has declined steadily from 90% in 1980 to 44% in 1985. Amorphous silicon and polycrystalline silicon have become more prevalent and in l985 accounted for 34% and 20% of the market respectively.

The manufacturing methods include the Czochralski process for producing crystalline wafers, which has been in use for several decades, and newer production methods such as thin film and ribbon processes. Photovoltaics is presently a highly dynamic technology, with materials and fabrication techniques constantly evolving, in response to the search for less costly and more efficient systems. A description of the principal types of PV cells follows.

Single-Crystal Silicon

Single-crystal silicon is a well-established technology, with cells that tend to be stable and relatively efficient. Its main drawback is the high cost of manufacturing.

The first step in producing silicon solar cells is the furnace reduction of silicon ores into raw elemental silicon which is then purified so there is less than one non-silicon atom per billion. The purified silicon is then re-melted at high temperature in a quartz crucible and a crystal is slowly drawn to form a large, cylindrical ingot, in a procedure known as the Czochralski process. The crystal is then cut into thin, round single crystal wafers of about 0.025 cm thick using a diamond saw. Unfortunately, half of the expensive silicon crystal is turned into useless dust in the slicing process. The crystal must contain traces of two materials in order to generate electricity. One, usually boron, is added to the molten silicon before it is crystallized. The other material, usually phosphorous, is later diffused into one side of each wafer, forming a barrier of electric charge between the two parts of the solar cells that direct the flow of electrons. Following this p-n junction formation, metal contacts are attached to the front and back of the wafer, and a glass or plastic encapsulant is added for protection.

Polycrystalline Silicon Cell

Polycrystalline silicon ingots with grain sizes of several mm can be produced by a casting process which is less expensive than the single crystalline process. The waste of material during the wafering process is still great, but the area factor in the modules is higher due to the rectangular shape of the multicrystalline wafers. Polycrystalline silicon is now also a well-established technology, with good stability and efficiencies only slightly lower than single-crystal silicon.

Polycrystalline Silicon Ribbon

An alternative to producing cystalline silicon wafers is the production of sheets or ribbons of polycrystalline silicon from the molten material, by-passing the expensive and wasteful slicing stage. Several ribbon technologies including edge-defined film-fed growth and dendritic web processes, are presently under investigation and their market share is expected to grow. The manufacturing process is complex, but the potential for high-speed production and high material utilization could offset the efficiencies which are somewhat lower than that of single-crystal cells.

Thin Films

In recent years researchers have been turning their attention to thin-film technology which offers the potential for making less expensive cells because they require less material and allow a more automated production process. Thin-film solar cells can be made from a variety of inexpensive semi-conductor materials which can be directly deposited on a substrate by techniques such as glow discharge, chemical vapor deposition and electrochemical deposition. The materials need be only a few microns in thickness or 1/10 as thick as conventional cells. Semi-conductor materials being investigated for use as thin films include amorphous silicon, copper indium diselenide, gallium arsenide, and cadmium telluride.

Amorphous Silicon Thin Films. The amorphous silicon solar cell is the most advanced of the thin-film technologies. It can be manufactured cheaply and easily and is well suited to automated mass production. These cells have a much higher absorption coefficient than single-crystal silicon, and the nature of their band-gap requires only a 1 μm thick active layer compared to several 100 μm or more for single-crystal silicon.

The main challenge is to obtain high efficiencies from a material that lacks the inherent photovoltaic properties of crystalline silicon. The other major problem associated with amorphous silicon cells is light-induced power degradation, known as the "Staebler-Wronski Effect", which can cause cells to loose 15% of their power output. The stability problem is most pronounced when the cell band-gap is optimized for natural lighting, and therefore, cells designed for interior use (e.g. in calculators) are less affected.

Non-silicon Thin Films. One of the earliest thin-film solar cells developed was the copper sulfide/cadmium sulfide cell. These cells have encountered stability problems, however, which have caused interest to shift to newer, potentially more effective materials. Other materials being investigated which are believed to have significant potential as thin-film cells are copper indium diselenide ($CuInSe_2$), cadmium telluride (CdTe), and gallium arsenide (GaAs). Copper indium diselenide offers a number of advantages as a thin-film material including relatively high efficiency and good stability over extended exposure to sunlight. Gallium arsenide crystal cells have achieved the highest efficiency of any PV material but are quite costly. The development of lower cost thin-film GaAs cells is thus of great interest. Dozens of other chemical alloys are also being studied by government and industry-funded researchers.

Multi-Junction Thin-Film Cells. Many of the thin-film materials are being investigated for use in multi-junction or tandem cells. Interest in multi-junction thin-film cells increased with the finding that stacking two *different* cell materials made it possible to take advantage of different sunlight absorption characteristics, capturing a broader spectrum of light and increasing cell efficiency. It was also discovered that stacking layers of amorphous silicon cells increased their stability. Multi-junction cell efficiencies in the 20%-25% range, or even up to 30%, are considered possible.

Concentrator Cells

An alternative approach for improving the output of photovoltaics is an optical concentrator system which increases the amount of sunlight striking a solar cell by ten to a thousand times. Concentrator technology attempts to circumvent the high cost of regular PV cells by covering the collector area with inexpensive concentrating optics and using a smaller amount of higher efficiency (15%) concentrator cells (See Figure 19). Laboratory cells have achieved efficiencies as high as 27% at concentration. Special plastic lenses (Fresnel lenses) focus the sunlight on to smaller cells (of 1 cm² or less). Mechanical trackers are required to allow the solar modules to maintain an optimal angle to the sun throughout the day.

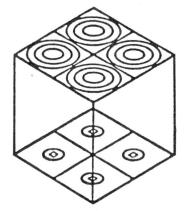

FIGURE 19
CONCENTRATOR CELL MODULE

FOCUSING LENSES

**HIGH EFFICIENCY PV
CONCENTRATOR CELLS**

There are trade-offs in using concentrating and tracker systems, however. Concentrator cells must be cooled with large passive heat sinks or actively cooled with fluids so that heat build-up will not impair efficiency. The durability of concentrating lenses currently lags behind flat-plate modules. The additional cost of the suntracking system is also a negative factor. Most importantly, the effective application of concentrator systems may be limited to geographic locations with abundant direct solar radiation, such as desert areas and certain clear, dry Mediterranean regions.

The efficiency of a PV system is directly related to cell efficiency, but certain losses will occur at the system level. Overall system efficiency is the product of the module efficiency and BOS efficiency. Using present silicon single-crystal cells (about 14% efficient), and projected low-cost flat-plate module construction, an average annual module efficiency of 11.2% is expected. This is 20% less than the basic cell efficiency because it includes optical losses in the module, the module packing factor, and the effect of an average annual cell temperature which is 20°C higher than laboratory rating temperature. Obviously, with 16% cells, module efficiency would be significantly increased, allowing smaller array fields for the same power.

BOS efficiency further affects system output. Wiring and module-to-module mismatch account for another 2% loss, and power conditioning accounts for about a 10% loss in energy. Therefore, only 99 W/m² or 9.9% of the 1 000 W/m² of incident sunlight on the module reaches the utility or the load.

III. APPLICATIONS

Photovoltaic applications can be divided into the following categories:

— Consumer products (0.001 - 1.0 W): The use of photovoltaics in specialty or consumer products such as watches, calculators, and other small electronic devices.

— Space applications (15 - 20 000 W): Systems that supply the primary or supplementary power needs of satellites and space missions.

— Stand Alone or Remote Systems (100 W-200 kW): Power systems for private residences, commercial establishments and villages that are not connected to an electric power grid. This category also includes power for remote microwave repeaters, signaling devices, cathodic protection, ocean buoy flashers and field telephones.

— Grid-Connected Residences (1-10 kW): Systems powering single and multiple family residence that supplement grid-connected electricity or sell excess power to a utility.

— Grid-Connected Intermediate (10-300 kW): Systems designed to provide the supplementary or primary electrical needs of commercial, industrial, or institutional facilities, with excess power being sold to a utility.

— Large Central Station (300 kW-1 MW or greater): electric utility-scale systems with output fed directly into the utility grid.

Pioneered by a few Japanese companies, the consumer specialty market has grown since introduction in 1979 to 6.7 MW in 1985 and probably now represents a major PV market segment. Using tiny PV arrays, these products can operate effectively without electric cords or batteries with the light available in most rooms. Because power requirements are so small, the solar array adds little to the overall production cost.

Aside from the consumer product market, PV power systems are most commonly used today for specialized applications in which their relatively high cost can be justified. These applications usually have some or all of the following characteristics:

— low power demand;

— expensive or unavailable power;

— government incentives to use solar energy.

These characteristics are found particularly in the remote, stand-alone power systems serving discrete energy demands. In these applications, PV power systems can often be the least expensive alternative because of the high cost of long-distance electricity distribution lines or other on-site power supply. PV systems have operated efficiently and reliably in such applications as microwave repeater stations, navigational buoys and isolated residences. The performance and economic viability of such systems can be improved and the load requirements better met if photovoltaic modules are used in combination with other electricity-generating renewable energy technologies with different supply characteristics, e.g. small wind energy conversion systems. Figure 20 shows an example of a solar/wind electricity-generating system for a remote application, supplying power to a special FM radio service. This market is relatively limited in size compared to other markets since the total worldwide power requirement in these applications is not likely to exceed a few tens of gigawatts, but it can be a good "seed" market for PV technology.

FIGURE 20
COMBINED SOLAR (PV)/WIND ENERGY ELECTRICITY GENERATING SYSTEM SUPPLYNG POWER TO A FM RADIO INSTALLATION AT A REMOTE LOCATION IN GERMANY

Source: AEG-Telefunken

A potentially larger market, but one still justifying a premium cost, is supplying village power, especially in developing countries. Many believe this will be the most important role for photovoltaics in the next decade since vast areas of the world are still not hooked up to utility lines and extension of the centralized grid system can be prohibitively expensive.

Given that two million diesel systems are sold worldwide each year and the fact that PV systems are often more reliable and are becoming cost-competitive, the potential market for PV systems in developing countries is clearly enormous. On the other hand, many of these people live in great poverty and providing them even with low-cost solar systems would be a great challenge.

Regarding the residential market, it is estimated that 15 000 homes worldwide now rely on solar electricity. Most of these have energy-efficient, 12-volt DC appliances so that the solar electricity can be used directly. For example, many residents of Monhegon Island, Maine, USA, utilize PV systems of a few hundred watts each to power a refrigerator, lights, CB radio, television, and vacuum cleaner. The total cost of such a system is currently between US $2 000 and US $5 000. Back-up power is supplied by batteries and/or a diesel generator. However, such systems do not provide the massive quantities of on-demand power to which many people are accustomed, although in remote areas these systems can be cheaper and require less maintenance than the alternatives. Other residential systems with greater power capacity, both grid-connected and stand-alone, have been built. They have been technical successes, but they are currently much too expensive for the mass housing market. Only if, as many analysts expect, solar module prices fall to US $1/$W_p$ and standardized assembly-line production brings the total system cost to less than US $3/$W_p$, then residential PV systems could become a viable alternative in some places.

A number of PV installations connected to centralized utility electrical grids have been built by utilities or with government funding, these installations are either primarily experimental or particularly strong financial incentives were available.

IV. STATUS AND ACCOMPLISHMENTS

Status in 1974

Between 1958 and 1974, the United States space programme created a small but steady demand of about 80 kW of power per year after it was found that the cells worked very well in space. Cell efficiency and durability were increased, and by 1970, costs had been reduced to between US $100/W-US $200/W. Nevertheless, since cost mattered little in the space programme where relatively few cells were used, photovoltaics were still at least fifty times more expensive than conventional electricity generation technologies, and few other applications were found.

With the dramatic rise in fossil fuel prices in the mid-seventies, interest in PV applications was renewed, and in 1975 terrestrial applications of solar cells outpaced space applications for the first time. The higher oil prices served as an impetus to the establishment of R&D programmes in several European countries, Japan, and the United States, which were aimed at increasing the efficiency of PV systems and reducing their cost. About a dozen companies were conducting research or manufacturing cells by the mid-seventies, and the annual production was at the level of a few hundred kW_p.

Technical Advances

Great strides have been made in photovoltaic technology as a result of substantial public and private sector research and development efforts over the past ten or twelve years. Module efficiencies have improved by a factor of two, materials promising even greater performance have been developed, and module costs have dropped from US $20/W_p$ in 1977 to US $5-8/W_p$ in 1985. Table 11 shows the increases in the efficiency which have been realized for some key cell types in the past few years. Government programmes begun in the mid-seventies have been instrumental in these impressive gains. The United States Government has spent approximately US $800 million on photovoltaics in the past decade, and the European nations and Japan have collectively spent another several hundred million dollars.

Outside the United States, the largest photovoltaics R&D programmes are in France, Germany, Italy and Japan. Other countries with smaller programmes include Australia, Belgium, Brazil, Canada, India, Mexico, the Netherlands, the Peoples' Republic of China, the Soviet Union, Spain, Sweden and the United Kingdom. Worldwide, photovoltaics has had one of the largest R&D budget of any renewable energy source.

Government funds have supported R&D widely from basic laboratory research to large demonstration projects, mostly through contracts with universities and private industry. Many private companies have also funded their own research programmes.

Research Areas

Research efforts have focussed on all aspects of the technology from basic materials to module efficiency and wiring. Cost reduction has been given especially high priority, and three basic strategies have been employed to address this issue. The first was the development of inexpensive, automated methods of manufacturing crystalline silicon cells, the primary material technology since the mid-fifties. The second was to develop alternative cell materials that are cheaper to produce. The third approach was to design systems that concentrate sunlight on the cells, thereby lowering the cost of the electricity produced even when using more expensive cell materials. Some of the important achievements of the last ten years of research are summarized below.

Single-Crystal Silicon. Improvements in material quality, production automation, and conversion efficiency have resulted in a much lower cost per watt. Single crystal PV cells now have production efficiencies in the 12%-14% range, while maximum laboratory efficiences of 21% have been achieved. Module life expectancy is over ten years.

Polycrystalline Silicon. This technology is catching up to single crystal cells in performance with 18% laboratory conversion efficiency and 11%-12% for commercially-produced cells.

Silicon Ribbon. Ribbon cell technology has been pursued because the process results in less material waste than the wafer slicing process and has the potential for high-speed manufacturing. Laboratory efficiencies of 17% and production cell efficiencies of 10%-13% have recently been achieved.

Amorphous Silicon Thin Films. Major research efforts have increased the laboratory efficiency of amorphous silicon from 4% to 13%. Light-induced power degradation (Staebler-Wronski effect) is still a problem but is gradually being overcome.

Other Thin-Film Technologies. Some of the most exciting developments have been in other thin-film materials and in combinations of different types of thin films in tandem or multi-junction cells (see below). Copper indium diselenide has been discovered to have relatively high efficiency (12% has been achieved) combined with good stability. Moreover, less expensive production processes such as electroplating rather than chemical vapor deposition appear feasible. Single-crystal cells currently of gallium arsenide (GaAs) have exceeded 22% efficiency for one sun and 26% under concentration. They are still considered too expensive, however, for terrestrial applications; therefore, considerable effort is being expended in the development of thin-film GaAs cells.

Multi-Junction Thin Films. Potential cell efficiencies in the 20%-25% range are considered possible for multi-junction cells. Presently, the highest efficiencies achieved have been in the 12%-13% range. Cadmium sulfide/cadmium telluride cells made by low-cost screen printing technology have eliminated the instability problem and have realised a conversion efficiency of 12.8% with small active areas in Japan. Gallium arsenide appears to offer the greatest potential in terms of tandem cell efficiency. Researchers hope to achieve efficiencies as high as 30% with a combination of GaAs and $CuInSe_2$.

Table 11 demonstrates some of the impressive gains that have been made in cell efficiency as a result of photovoltaic research programmes.

TABLE 11
PROGRESS IN PHOTOVOLTAIC CELL EFFICIENCY, 1982-86
(Laboratory Cells)

	1982 %	1983 %	1986 %
Single-Crystal Silicon	12	16	19
Polycrystalline Silicon	10.5	14	18
Silicon Ribbon (EFG)	12	12	15
Amorphous Silicon	5	7	12
Gallium Arsenide	15-17	23	24
Copper Indium Diselenide	10	11	12.5
Dendritic Web	14	15	16

Sources: 1) Jeffry Erickson and Carlo La Porta. *Energy Innovation: Development and Status of the Renewable Energy Industries.* Washington, D.C., Solar Energy Industries Association, pp. B1-B11.

2) "Alternative Sources of Energy", May 1986, pp.8-11.

Concentrator Systems. Prototype GaAs concentrator cells have achieved 26% efficiency at a concentration factor of 750, but these cells are not being produced on a regular commercial basis for terrestrial applications. There is more experience with single-crystal silicon concentrators since several test installations have been built. When fabricated under particularly careful conditions and high performance cells from a larger production run are used, these cells have a demonstrated efficiency of 20% at a concentration factor of 150. Incorporated in an experimental module at standard test conditions, 17% efficiency has been achieved.

An advanced approach is to design a single crystal silicon cell which is specially optimized around the conditions of high concentration with a point contact geometry of multiple p-n junctions, all on the back of the cell. A theoretical efficiency of greater than 30% has been predicted for such a design and recent measurement of such a device at Stanford University indicates a 27% efficiency at 100 suns concentration. Such promising results to date suggest that it is reasonable to expect crystalline silicon concentrator modules to meet the efficiency target of 22%-27% with further technical development.

Balance of System Components. In the past several years, engineers have successfully improved power conditioners, cut support structure costs, and simplified wiring and installation methods. Researchers are also looking into alternatives to lead-acid batteries such as sodium-sulfur and nickel-zinc batteries, but progress is slow.

Pilot Plants

In addition to laboratory research on cell materials, production and PV components, a number of important pilot projects have been constructed which are providing key data on system performance and reliability, and on plant operation. These projects have been funded both by governments and private industry.

Four large MW-class PV pilot plants are in operation, three in California, United States and one in Japan. Information on these and a number of other plants above 100 kW is found in Table 12. The Commission of the European Communities has provided partial funding for 15 pilot projects in its member countries. These installations range in size from 30-300 kW (see Table 13). Spain's experimental 100 kW PV plant at San Augustin de Guadalix became operational in March 1985. The plant is divided into two grid-connected units of 40 kW and 50 kW and an unconnected 10 kW array. Within the framework of the Greek five-year programme (1984-1988) thirteen very small islands will be electrified by small PV units for a total capacity of 60 kW. Another PV plant is under construction for irrigation on one island.

Commercial Status

The photovoltaics industry has grown rapidly in the past few years as new markets have been identified and developed. New and increasingly better products are being brought to the marketplace, and commercial sales, while still miniscule compared to other energy sources, have until recently been expanding at a 50% or better rate annually. Global production has increased from about 1 MW of PV systems in 1978 to 24 MW in 1985. Estimated sales have risen from US $85 million in 1980 to between US $350 and US $400 million in 1984. As a result, costs of installed systems and delivered power are declining steadily.

Description of the Industry. A modest-size photovoltaic industry has emerged in Australia, Belgium, Canada, France, Germany, Italy, the Netherlands, Japan and the United States. More than 100 countries around the world are currently producing photovoltaic cells and modules for commercial use. As of 1984, twenty-three United States companies were manufacturing systems, while about fifty companies were involved in either sales or research. Four of the top seven United States PV manufacturing firms are owned by major oil companies. Twelve companies are producing PV modules in Europe. The photovoltaic industry there has been dominated by major electronics companies. Fourteen Japanese companies manufacture solar cells. As in Europe, the electronics industry in Japan is heavily involved in photovoltaics. Three Australian companies are manufacturing solar cells. Approximately ten companies are producing PV devices on a small scale in developing countries.

TABLE 12
LARGE PHOTOVOLTAIC PILOT PLANTS (More than 100 kW)

Project (& Sponsor)	Location	Power Output (kW)	Actual or Expected Completion Date
Community College (Department of Energy)	Arkansas	245	1981
SOLARAS (Saudi & U.S. Governments)	Saudi Arabia	350	1981
Sky Harbor Airport (Department of Energy)	Arizona	225	1982
Solar Breeder (Solarex)	Maryland	200	1982
Lugo Station/Hesperia (ARCO Solar/Southern California Edison)	California	1 000	1982
Pellworm Island (European Economic Communities)	Germany	300	1984
Georgetown University Intercultural Center (Department of Energy)	Washington, D.C.	300	1984
Carrisa Plain (ARCO Solar/Pacific Gas & Electric)	California	6 400	1983
Rancho Seco (Sacramento Municipal Utility District)	California	2 000	1984
NEDO/Tsukuba (Japan MITI)	Japan	200	1985
NEDO School Project/ Ichihara (Japan MITI)	Japan	200	1986
Solar Research Project (Alabama Power Company)	Alabama	110	1986
NEDO utility system/Saijo (Japan MITI)	Japan	1 000	1986
Delphos (Italian Government)	Italy	300	1987

Although the United States pioneered the production of photovoltaics and as recently as 1980 produced 80% of the world's solar cells, the global market has become much more competitive. The United States market share had consequently fallen below 50% by 1985. This decrease was due in part to a fall-off in central utility applications. At the same time, Japan's market share rose because of the continued popularity of small consumer goods such as calculators and watches. In 1984, this market accounted for 86% of Japanese shipments.

TABLE 13

**COMMISSION OF THE EUROPEAN COMMUNITIES
PHOTOVOLTAIC PILOT PROJECTS**

kW_p	Location	Application
50	Crete Island, Greece	Village power
50	Mont Bouquet, France	TV transmission
50	Fota Island, Ireland	Dairy farm
35	Kaw, French Guiana	Village power
50	Terschelling Island, The Netherlands	School power
44	Corsica Island, France	Village power
80	Alicudi Island, Italy	Village power
63	Chevetogne, Belgium	Swimming pool
50	Nice, France	Airport control
65	Tremiti Island, Italy	Desalination
45	Giglio Island, Italy	Water cooling
30	Hoboken, Belgium	Hydrogen output
30	Marchwood, England	Utility network
100	Kythnos Island, Greece	Island power
300	Pellworm Island, Germany	Island power

Market Penetration

Single-crystal silicon, with a 1985 production level of 10.8 MW_e and representing a world market share of 44%, remains the mainstay of the photovoltaics industry, (sales of these cells comprised 62% of the United States PV sales). Polycrystalline silicon production amounted to 4.9 MW_e or 20% of the world market. Amorphous silicon sales are now in second place with a 1985 production of 8.45 MW_e representing 34% of the current world market. Sales of silicon ribbon cells are still small, 0.15 MW_e, a 2% market share.

Estimates of world PV shipments and market share for 1983 through 1985 are given in Table 14. A range of figures is provided, reflecting divergence in statistics provided by different PV analysts. Almost all the experts agree on two notable trends: a decrease in total shipments for 1985 over the previous year and Japanese replacement of the United States as market leader. The decline in shipments is due in large part to a decline in government purchases. Battelle analysts are projecting a major increase for 1986 over their 1985 figure of 20.5 MW$_e$. They expect total shipments to rise to 25.0 MW$_e$ with large gains in both the specialty and stand-alone markets.

TABLE 14
**ESTIMATES OF WORLD ANNUAL PHOTOVOLTAIC SHIPMENTS
BY COUNTRY AND MARKET SHARE**

	Megawatts Shipped			Market Share		
	1983	1984	1985	1983 %	1984 %	1985 %
U.S.	9.3-13.1	8.5-15.0	7.5-10	60	46-54	37-39
Japan	3.8- 5.3	6.6- 8.9	6.4-11	21-24	28-36	33-41
Europe	2.2- 3.5	2.6- 4.0	3.7-4.5	14-17	14	17-19
Other*	0.3- 0.5	0.8- 1.0	1.8-1.5	1-2	3-4	5-9
World	15.6-22.4	18.5-28.9	19.4-27	———	———	———

* "Other" is essentially two photovoltaic manufacturers in India and Brazil.

Sources: *Photovoltaic Industry Progress* (Battelle Memorial Institute, August 1986), Photovoltaics Insider's Report, PV News, and Strategies Unlimited.

A close examination of the photovoltaic market by end-use (see Table 15) reveals a marked decrease in central station and grid intermediate markets which had accounted for 68% of United States PV shipments in 1983. The future of this market in the United States is likely to depend on improvements in cell efficiencies, system economics, utility needs for additional power capacity, and the cost of conventional energy.

In contrast, the world market for speciality or consumer products has risen significantly. Japan developed this market, particularly for calculators, in 1981 and has dominated it ever since. Over three-quarters of the solar cells manufactured in Japan are used on consumer electronic devices, and amorphous silicon cells are the cells most commonly used for these devices. This has been a major boost to the Japanese photovoltaics industry because it is a relatively high-priced market with the small individual cells for consumer products commanding up to US $50/W. Although this market is quite successful and lucrative, it does not represent a serious contribution to world energy needs.

The market share of stand-alone applications has risen for the past three years, regaining its earlier dominance over the other end-use markets. Further gains are projected since stand-alone systems are more cost-effective than other small power sources in many remote applications. It is estimated that between 5 000 and 10 000 remote PV systems rated one kW or less are sold in the United States each year. Some isolated homes and communities have turned to PV as a power source, because of the high cost of connecting to a central utility. For example, thousand of vacation homes

TABLE 15
ESTIMATED WORLD PV SHIPMENT TO END-USE MARKET SECTORS
(Capacity and Market Share)

	1981		1982		1983		1984		1985		1986	
	(MWe)	(%)	(MWe)	(%)	(MWe)	(%)	(MWe)	(%)	(MWe)	(%)	(MWe)	(%)
Speciality	0.8	17	1.2	15	4.1	19	6.2	27	6.7	33	8.9	36
Stand-Alone	3.6	76	3.9	52	7.7	36	10.7	46	11.0	53	12.9	52
Grid Residential	0.1	1	0.1	2	<.1	1	0.3	<.1	1.0	1.5	0.4	2
Grid Intermediate	0.3	6	0.5	7	1.0	5	0.6	3	0.6	3	0.6	2
Central Station	0.0	0	1.8	24	8.7	40	5.7	24	1.9	9	2.3	9
Total	4.8	100	7.5	100	21.5	100*	23.2	100*	20.5	100*	25.0	100*

* Total rounded.

Source: J.A. Dirks, S.A. Smith, and R.L. Watts. *Photovoltaic Industry Progress Through mid-1986*. Richland, WA, Batelle Memorial Institute, Pacific Northwest Laboratory, August 1986.

in Norway utilize small PV systems. A recent development dedicated to these markets and using single-crystel or polycristalline photovoltaic modules, mounted on and integrated into the roof tiles, was achieved by a German company and is shown in Figure 21.

FIGURE 21
"Forschungshaus Remscheid", the world's first residence with photovoltaic roof tiles. PV cells are visually and technically integrated intro roof architecture. Optionally 432 PV tiles at 8.4 W_p (polycrystalline cells) or 9.6 W_p (single-crystal cells) produce 3.6 kW_p (4.1kW_p) and can supply up to 7800 kWh/year in moderate latitudes (Germany).

Source: BMC Solartechnik GmbH.

Developing countries represent a promising market, especially in remote villages where access to electricity is restricted. The United Nations has estimated that over two million villages worldwide do not have electricity. To help stimulate this market, a number of governments have sponsored demonstration projects in developing countries. It is estimated that, by the early 1990s, 46 MW of photovoltaics will be located in remote areas, as compared to 4 MW in 1983.

Finally, it is interesting to note the breakdown of shipments for 1985 by cell type, which is indicated in Table 16. (Once again, it should be noted that industry analysts differ in their estimates of PV production and sales since methods of determining these figures vary and it is difficult to obtain precise numbers.)

Costs are the most critical factor in determining commercialization status. The lowest per unit costs are obtained for large (more than 500 kW) utility-connected systems. These systems consist only of the PV modules or panels, the array field (site preparation, structure, wiring), and the power processing hardware.

TABLE 16

WORLD PV SHIPMENTS BY CELL TYPE BY COUNTRY AND CELL TYPE-1985

(in MW$_e$)

CELL TYPE	U.S.	Japan	Europe	Other	Total
Single-Crystal Silicon	5.3	2.5	1.6	1.4	10.8
Polycrystalline Silicon	2.4	0.4	2.1	0	4.9
Amorphous Silicon	0.6	7.8	0	0	8.4
Concentrating	0.05	0	0	0	. 5
Ribbon & Other	0.15	0	0	0	.15
Total	8.50	10.7	3.7	1.4	24.3

Source: Photovoltaic News, February 1986

For these large systems, module costs recently have been typically about US $5.00/W$_p$. The array field costs for fixed flat plate collectors are in the range of US $50-US $80/m^2. For a module of 12% efficiency, this adds US $.40/W$_p$ to US $.70/W$_p$ to the system cost. The power processing hardware currently adds another US $.60/W, for a total hardware cost of about US $5.50/W - US $5.80/W. Indirect costs, such as engineering fees, interest during construction and contingencies, raise today's large system price to US $7.00 - US $8.00/W$_p$. This corresponds to energy costs of US $.30-US $.35/kWh. Operation and maintenance costs would be small over a plant's assumed thirty-year lifetime, increasing these energy costs by less than 5%.

The unit cost of smaller systems will typically be higher because of loss of economies of scale in module purchases and the increasing impact of fixed costs on site preparation and power processing hardware. For stand-alone systems, the use of battery storage is another major cause of system cost increase.

An example of the upper extreme of system costs would be a 1 kW system located in a sunny location, meeting a 5 kWh/day load. To assure the capability of the system to meet the load during inclement weather, three days of energy storage (or 15 kWh) will be incorporated. This would add US $3 700 to the system cost, assuming US $250/kWh. The power processing hardware, depending on its sophistication and the type of loads to be met, can add another US $2 000 to the system cost. The structure and wiring would cost about US $1 000. If modules were purchased in large blocks to supply numerous systems, it would be possible to achieve module costs close to those for large systems. Assuming this were done, the total price for the example system would be about US $14 000, including indirect costs. The energy storage contributes over 33% of this cost in the form of batteries and charge control hardware.

V.   ENVIRONMENTAL CONSIDERATIONS

During their operation, photovoltaic systems discharge no gas or liquid emissions or heat and are therefore considered ecologically benign. The PV production process, however, can involve hazardous substances as in other semi-conductor industries.

Fabrication of thin-film cells requires large quantities of gases, some of which are toxic (AsH_3, PH_3, SiF_4). Accidental release of hazardous air pollutants could present a risk to workers or persons living close to the facility. Care must therefore be taken in handling and disposal of these gases. Under normal conditions, the quantities of pollutants generated from PV plants are small. The most commonly used control system is wet scrubbing. Commercially available multi-stage scrubbers can be at least 98% effective in removing emissions from effluent streams.

Solid wastes resulting from the manufacturing process vary in quantity and quality. Some are toxic, such as cadmium and arsenic compounds and require careful handling and disposal in controlled landfills.

The possibility of fire from rooftop PV arrays is another potential danger. Fires can be caused by short circuits or by spontaneous combustion due to heat build-up in dead air spaces. There is also a risk of electrical shock to installers if normal safety precautions are not taken. Safeguards for PV components can be built into PV systems, such as the diodes, wiring systems, and mounting frames to prevent electric shock and fire hazards as have been specified by the Underwriters Laboratory in the United States.

VI.   TECHNOLOGY OUTLOOK

Work Underway and Technical Prospects

Enormous progress has been made in photovoltaic technology, but further improvements in cell efficiencies and system cost are essential. Researchers have a good idea as to the achievable limits of cell efficiencies and are working hard to achieve those levels (See Table 17). In many cases, laboratory efficiencies very close to these theoretical limits have already been attained.

In single crystal silicon, efforts are underway to further reduce the production cost of purified silicon, to develop high speed crystal pullers and wafer slicing techniques, and to improve overall design and efficiency of modules. It is hoped that module efficiency will reach 17%. Researchers are also attempting to produce polycrystalline cells from a less pure form of silicon that costs only a fifth as much as the semi-conductor-grade silicon that is now in use.

TABLE 17
POTENTIAL CELL EFFICIENCIES

Cell Type	Achievable Efficiencies %
Single-Crystal Silicon	25
Single-Crystal Silicon Concentrator	29
Polycrystal Silicon	22
Ribbon Silicon	20-25
Amorphous Silicon	15-18
Multi-Junction Thin Films	
Amorphous Si/CuInSe$_2$	20-25
CdTe/CuInSe$_2$	25-30
GaAs/CuInSe$_2$	25-30
Amorphous Si Stacked	18-20

Source: Alternative Sources of Energy, May 1986, p. 10, and personal communication with Andrew Krantz, U.S. Department of Energy, September 1986.

New and advanced photovoltaic materials and deposition processes are being intensively investigated for thin-film applications as are combinations of materials for multi-junction or tandem cells. Aside from amorphous silicon, neither single-junction nor multi-junction thin-film cells are being produced comercially as yet.

It is likely that a 15% efficiency for amorphous silicon (aSi) thin-film cells can be reached fairly soon which would assure the viability of amorphous silicon for power applications in addition to its widespread use in consumer speciality products. In fact, with tandem devices, developers expect 15%-18% efficiencies. Moreover, with a potentially lower price per W_p than that of other PV materials, aSi modules are considered by some to have the greatest promise for power application, if stability and efficiencies can be improved. Degradation of 15% in aSi production cells is still common, and manufacturers are addressing this problem.

Research continues on PV concentrators, and they have achieved excellent laboratory efficiencies; however the geographic limitations and cost of the tracking mechanism have been noted earlier in this discussion. Special systems able to concentrate diffuse radiation, such as the fluorescent concentrator, have not yet proved their feasibility because of continued low efficiencies.

The cost of labor is a large component of solar cell manufacturing costs because the manufacturing process is slow and labor-intensive. Development of new processing technologies and automated production methods could be an important factor in achieving lower-cost cells.

Module and system lifetimes and operating reliability have been improved. Therefore, studies being conducted of cell and module durablity under environmental stress, degradation studies of cells and encapsulation materials, and accelerated aging tests will be quite valuable.

Efforts are continuing to lower balance-of-system costs which represent a major portion (about 30%) of the cost of a PV system. Research is being conducted on less expensive batteries, but advances have been slow. The cost of the major materials used in solar electric installations, such as structural steel, concrete, and copper wire, are also likely to remain relatively stable. However, in the last few years, engineers have successfully improved the efficiency of power conditioners, cut support structure costs, and simplified wiring and installation methods.

There is also insufficient knowledge regarding optimum system configuration, performance and operation because of the relatively limited number of pilot plants. The construction of additional systems would help provide essential data on performance and operating experience.

Finally, the current state of knowledge of the solar resource in many countries must be improved for optimum design and siting of PV plants, especially grid-connected systems. Before utilities can integrate PV into central plants they would require minute-by-minute understanding of insolation patterns at proposed sites and considerably more data, specific to the needs of PV system designers.

Outlook

The outlook for photovoltaic technology is quite promising in view of the conversion efficiency improvements which have been achieved, the expanded system lifetime, and continuous efforts on reducing system costs. The new thin-film and multi-junction technologies are likely to gradually dominate the market over the coming decade; although the single-crystal and polycrystalline cells will probably continue to be produced and sold for some time. Systems are expected to be relatively trouble-free, with low operating and maintenance costs. Furthermore, based on field experience to date, a system life expectancy of thirty years appears to be a reasonable expectation.

Most researchers seem in agreement that the US $1.00/$W_p$ price for modules is well within reach. The wide diversity of approaches being taken in pursuing that goal helps to ensure that it will be accomplished. The United States Department of Energy sees total system costs being reduced from the present level of over US $9 000/kW to US $1 400/kW-US $1 900/kW in the late 1990s, given the expected reductions in cell costs and balance-of-system costs.

For the needed cost reductions to be achieved, production will have to be automated, and cells will have to be produced on a large-scale, continuous basis. Some experts think that a minimum annual plant production of 100 MW would be required to bring about a manufacturing cost which would allow competition with conventional electricity generation costs.

In the United States, the extension of the solar tax credits for business may help to promote continued growth of PV utilization; although the effect of the elimination of the residential solar tax credits (which occurred at the end of 1985) is expected to result in very slow growth in the residential market.

Increasingly utilities view PV as useful additions to their supply options. With many companies fearing excess capacity or suffering from the effects of multi-billion dollar nuclear projects, the small scale, modularity and short lead times of photovoltaic systems present an attractive option. A United States Office of Technology Assessment study expects grid-connected PV plants to be important in the United States and Japan in the 1990s.

Some applications are expected to become cost-effective sooner than central PV power plants. Rapid growth is expected in remote, stand-alone applications in the Third World, where PV systems should begin replacing diesel generators as module prices fall to US $3-US $4/$W_p$. The use of hybrid PV/diesel power systems is also a possibility. But this scenario could be dependent on improved economic conditions in those countries, assistance from international aid agencies, and oil prices.

A number of optimistic projections have been made regarding increased use of photovoltaics. One industry analyst projects that sales will rise from 24 MW in 1985 to 500 MW in 1990. That would make photovoltaics a US $2 billion per year business by that date. A major 1984 study by Frost and Sullivan Inc., forecasted that total installed generating capacity will reach 4 500 MW in the year 2000.

Projections are far from certainties, but it is certain that photovoltaics is one of the success stories of the past decade's energy R&D, and its future is likely to be very bright.

VII. **MAJOR FINDINGS**

Photovoltaics is a technology under development which is capable of supplying either centralized or decentralized electrical energy.

The key to module improvements are advances in basic understanding of materials and formation of the cell and/or thin-film materials.

Concentrator cells have demonstrated very high efficiencies, but their usefulness may be limited to areas with predominantly clear skies. Also, the need for a tracking mechanism adds to the cost of the system.

To date, crystal silicon PV cells have exhibited higher efficiencies than amorphous silicon and non-silicon thin-film cells. However, single and multi-junction thin-film cells combine the advantages of low material use, less expensive production processes, and the potential for much higher efficiencies. Many experts believe that thin-film cells will dominate in the future. The most promising thin-film materials appear to be stacked amorphous silicon, copper indium diselenide and cadmium telluride.

While care has to be taken in the production process with certain hazardous materials, PV systems are ecologically benign during operation.

In 1985, the annual worldwide production of PV cells was between 20-24 MW, up from about 1 MW in 1978. Japan, the United States, and a few European countries, including Germany, France, and Italy, dominate the market.

The primary markets to date have been for consumer products, small stand-alone systems such as microwave repeaters, field telephones, navigational buoys, and small residential systems. The markets for Third World electrification and grid-connected systems have had only limited penetration thus far, but the potential is great if further cost reductions are achieved.

Increasing system lifetime improves cost-effectiveness. Today's data indicates a very favorable 30-year life-time.

For larger systems (more than 500 kW), module costs are typically about US $4.50/$W_p$ today with a 12% module efficiency. A US $1/$W_p$ goal for large systems appears to be a reasonable expectation for the next decade.

While primary effort must be placed on reducing the cost of modules, emphasis must also continue on reduction of balance-of-system costs and the costs of construction as well as the achievement of mass production since these are crucial to lower system costs.

VIII. BIBLIOGRAPHY

1. Harry K. Charles, Jr., *Handbook of Energy Technology and Economics,* John Wiley and Sons, New York, 1983, p. 663.

2. E. Berman, R.F. Reinoehl, J.C. Arnett and J.H. Caldwell, ARCO Solar, Inc. Large Scale Photovoltaic Energy System, report prepared for 1st International Photovoltaic Science and Engineering conference, Kobe, Japan, November, 7 pp.

3. A brochure made by New Energy Development Organization (NEDO), Japan, September 1984, 55 pp.

4. Yoshihiro Hamakawa, Osaka University, Japan. 1st International Photovoltaic Science and Engineering Conference and the backgrounds, Energy 1984, Vol. 17, Nov. 11, pp 29-35.

6.   J.G. Witwer, Electric Power Research Institute, Photovoltaics Power Systems Research Evaluation, Palo Alto, California, Dec. 1983.

7.   Department of Resources and Energy of Australia, A Review of the Photovoltaics Sub-Program, Canberra, Australia, 1985.

8.   New Energy Development Organization's brochure, September 1984, New Energy Information Volume 9, NEDO, 1983.

9.   The ARCO Solar Photovoltaic Power Plant, Carrisa Plain, Technical and Organizational Background Information, ARCO Solar Inc., November 1984.

10. The ARCO Solar Photovoltaic Power Plant, One MW-Hespera, California, Technical and Organizational Background Information, ARCO Solar Inc.

11. Solar Energy Intelligence Report

12. Financial Times, Energy Economist, April 1985 Photovoltaics move into the megawatt range.

13. OECD report on Environmental Impact of Renewables ENV/EN/85.1 2nd revision.

14. Alternative Sources of Energy, No. 81, May 1986, pp 8-15.

15. Alternative Sources of Energy, No. 78, February 1986, pp. 29-33, 51.

16. *Electricity from Sunlight: The Emergence of Photovataics,* Christopher Flavin for Solar Energy Research Institute, Dec. 1984.

17. *Annual Renewable Energy Technology Review, Progress Through 1984,* Renewable Energy Institute, 1986.

18. J.A. Dirks, S.A. Smith and R.L. Watts. *Photovoltaic Industry Progress Through Mid-1986,* Battelle Memorial Institute, Pacific Northwest Laboratory, August 1986, PNL-5954.

19. Personal communication from Gary Jones, Sandia National Laboratory, Sept. 4, 1986.

CHAPTER THREE: WIND ENERGY

A. NATURE OF THE WIND ENERGY SOURCE

Atmospheric Physics

Wind exists because solar radiation reaches the Earth's highly varied surface unevenly, creating temperature, density and pressure differences which cause air to move. On a global scale, these atmospheric currents work as an immense energy transfer medium. Cooler air from the poles replaces warm air rising from equatorial zones, thereby exchanging energy between these disparate climatic zones. Rotation of the planet further contributes to the establishment of semi-permanent, planetary-scale circulation patterns in the atmosphere.

Besides these major forcing agents, other factors such as topographical features alter wind energy distribution so that many exceptions to the global pattern exist, especially on a local scale. For example the different surface absorbance characteristics of land and water along a coastline can create winds. Valleys and mountains act similarly.

Resource Classification

Because wind speed is inconsistent, data must be averaged over time. Resource availability for a location is commonly expressed as a yearly average in meters per second or miles per hour. These averages may differ as much as 25% from year to year.

The level of kinetic energy in the wind is a function of the volume of mass (air), the density of that mass, and the velocity at which it travels. Assessment of wind resource potential begins by calculating power density which is expressed as power per unit area, e.g. W/m^2. The power that may be extracted from the wind is: 1) not affected by changes in air density except for extreme climatic locations and at high altitudes; 2) directly proportional to the area intercepted (a factor of two increase in the swept area doubles power output); 3) highly affected by velocity (wind speed in the power formula is a cubic function, thus a doubling of velocity increases power eight times).

Significant seasonal differences exist in wind resources in most areas. Generally wind speeds are higher in winter, peaking in spring. There are exceptions, such as California, in the United States, where summer winds are stronger. In any case, seasonal variations are extremely important because the cubic function that determines power output from a wind turbine means that energy potential can be significantly higher than an annual average would indicate if strong winds exist seasonally. There are also daily and hourly variations which are important to users like electric utilities that must adjust output of their power stations to meet variations in peak loads.

Many different methods have been used to classify wind resources. Table 18 illustrates one approach, a division of the wind resource into seven classifications, with data on power density and wind speed at two elevations for each. This table illustrates how power density increases rapidly as wind speed increases and how higher elevations improve potential energy capture.

TABLE 18

CLASSES OF WIND POWER DENSITY AT 10 AND 50 METERS ELEVATION

Height Above Ground	10 m (33 ft)		50 m (164 ft) (a)	
Wind Power Class	Wind Power Density W/m²	Speed (b) m/s (mph)	Wind Power Density, W/m²	Speed (b) m/s (mph)
1	100	4.4(0.8)	200	5.6(12.5)
2	150	5.1(11.5)	300	6.4(14.3)
3	200	5.6(12.5)	400	7.0(15.7)
4	250	6.0(13.4)	500	7.5(16.8)
5	300	6.4(14.3)	600	8.0(17.9)
6	400	7.0(15.7)	800	8.8(19.7)
7	1000	9.4(21.1)	2000	11.9(26.6)

(a) Vertical extrapolation of windspeed based on the 1/7 power law
(b) Mean windspeed is based on Rayleigh speed distribution of equivalent mean wind power density. Windspeed is for standard sea-level conditions.

* Prepared by Pacific Northwest Labs for the United States Department of Energy and United Nations World Meteorological Organization, in Technical Note No. 175, 1981.

Because wind resources vary in speed, direction and availability, measurement and documentation of the available resource is a crucial step for its ultimate development. Determining truly exploitable resources is a complicated process, however, involving not only the resource assessment, but evaluation of technology capability to match the resource. Research presented at the 1984 European Wind Energy Conference by MBB of Germany reported that 170 000 potential sites existed in the European Community for 100-meter wind systems, representing a total annual potential of 2×10^{12} kWh. This and other highly positive resource assessments indicate why many alternative energy programmes have assigned wind energy resource development significant priority.

Several countries have performed detailed wind resource assessments. In the United States, a compendium of twelve wind resource atlases has been published. The European Communities research programme will soon publish a wind atlas for the countries in the Community. Under sponsorship of the United Nations, the Pacific Northwest Laboratory in the United States has developed a worldwide resources map, of which the results are compiled in Figure 22.

FIGURE 22
ESTIMATED WORLD-WIDE DISTRIBUTION OF WIND ENERGY SOURCES

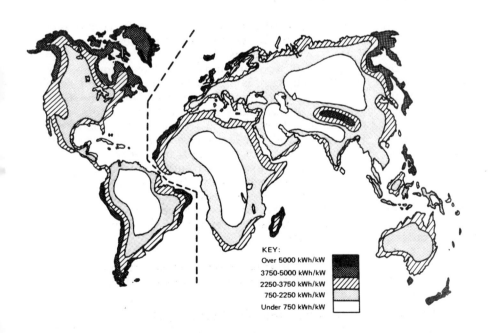

KEY:
Over 5000 kWh/kW
3750-5000 kWh/kW
2250-3750 kWh/kW
750-2250 kWh/kW
Under 750 kWh/kW

Source: United States Department of Energy

B. WIND ENERGY CONVERSION SYSTEMS

I. INTRODUCTION

Peoples in the Near East and Egypt developed the earliest wind systems to lift water or grind grain and by the thirteenth century their invention had begun to spread to Europe. Through the centuries its use expanded, and wind became important for providing mechanical power, producing electricity and pumping water, particularly in isolated locales. For example, by the early 20th century installed wind capacity for all applications in Denmark totalled 150 to 200 MWe, and Danish industry relied on wind power for one quarter of the energy it used. The use of wind declined everywhere to a negligible level with the development of oil and gas resources and rural electrification. From the 1930s, the only development occurring was the limited experimental work being carried out in a few countries - Denmark, England, France, Germany, the Soviet Union and the United States.

II. TECHNOLOGY DESCRIPTION

Modern wind machines generally come in two basic configurations: horizontal axis, and vertical axis. Horizontal axis machines may be divided into head-on and cross-wind systems. Vertical axis machines include panemone, Darrieus, giromill, Savonius, and others. Except for the Darrieus units, few have been the subject of comprehensive research and development programmes. Consequently, propeller horizontal-axis and Darrieus ("egg-beater") vertical-axis technologies have dominated technology and market development in the last decade. The aerodynamics of both types of machines is not well understood. For larger propeller machines sheer size increases their complexity. Per unit of swept area, propeller machines require less material and have less expensive rotors than vertical-axis units. Proponents of propeller equipment point out that they have lower cut-in speeds (the wind velocity at which the machine will self-start) and have a superior performance coefficient over a wider range of wind speeds. Darrieus machines have the advantage of having the drive system and generator located at the ground.

A wind machine intercepts flowing air and converts it to rotating shaft power. Horizontal and vertical axis configurations each have the same fundamental subsystems:

i.   The rotor, which is the assembly of blades, hub and shaft;

ii.   A drive train, which usually includes a gear box and electric generator, or mechanical drive;

iii.   Tower or mart, to support the rotor system;

iv.   Control system; and

v.   Balance of plant subsystems, including electrical interconnection, service facilities, supporting structures.

Early systems used the drag force of the wind to operate. Modern systems use aerodynamic lift across the propellers or blades.

Horizontal axis machines are the most common units manufactured, however, interest in Darrieus systems is found because they can accept the wind from any direction without having to yaw (turn). Small upwind horizontal axis systems usually employ a tail vane, which increases their weight, and downwind systems must adjust for the effect of their tower on the wind stream. Larger horizontal axis systems require micro-processor-controlled yaw systems, which increases their cost. Darrieus machines have their generator located near the ground, which facilitates installation and maintenance and simplifies structural wind loading concerns.

Up to 20 kilowatts in size, most horizontal-axis machines use three blades in an upwind design. As these machines increase in scale, two blades are preferred due to the substantial weight penalty. West German companies have developed counterweighted single blade prototype machines. Darrieus machines that have been introduced commercially have tended to be two-bladed. Blades may be fabricated of aluminum (cast or extruded), wood, and fibreglass plastic. Aluminum can be less costly in volume production. However, some manufacturers have found structural flaws in aluminum castings. Extruded aluminum blades have performed well to date on Darrieus systems. Wood blades are used in smaller machines because they have demonstrated excellent fatigue characteristics. Despite the fact that the potential exists for water penetration of wood or wood laminate blades in larger machines causing an imbalance in the rotor, a number of manufacturers do not consider it a serious disadvantage.

Commonly fitted to turbines up to 100 kW in size, fibreglass-reinforced polyester (FRP) blades have been chosen by many wind turbine manufacturers. The comparatively lighter weight FRP produces less stress on bearings and rotor hubs. Manufacturers must guard, however, against environmental degradation of FRP. A few turbine manufacturers have used steel blades, especially for larger turbines, more than 30 kW. Potential damage from corrosion must be weighed against the gains—ease of fabrication, strength, and lower cost.

Rotors on horizontal-axis systems may be mounted either up- or downwind. Very small systems are upwind because the tail vane provides a simple and low cost yaw (pointing) system. As systems grow larger, some manufacturers consider a downwind design less costly. Nonetheless, large Danish turbines and the MOD-2 units in the United States are upwind systems. Downwind turbines do require careful system design to mitigate the effects of tower interference with air flow and the blade ("tower shadow").

Wind turbine aerodynamic efficiency may be controlled by adjusting the angle of attack (pitch) of the air foil to the dynamic conditions. Manufacturers choose among fixed pitch, passive control, partial-span active control, and full-span active control. While variable pitch machines are more efficient than fixed-pitched machines, manufacturers have to consider possible penalties from increased system complexity. Variable pitch systems provide aerodynamic braking, thereby adding protection against overspeed.

Larger wind systems generally use synchronous AC generators and rotors designed to run at a constant speed. DC generators were common on the earliest small wind electric systems, taking advantage of their ruggedness, efficiency and ability to self-start. DC generators are more expensive, however, due to higher manufacturing and material costs (e.g. copper). In addition, DC generator brushes require periodic replacement. Since systems interconnected to utility power systems are most in favor, AC systems have some market advantage. For such an application, DC systems are supplied with DC/AC synchronous inverters.

Wind turbines are most often classified by their maximum rated (nameplate) power output, expressed in kilowatts. This practice acknowledges the importance of the electric utility market and provides a familiar reference for investors unfamiliar with the technology. A drawback of such a classification system is the dependence of wind power output on wind speed. Analysts evaluating equipment consider the area swept by the rotor a better indication of the potential energy output of a turbine. The International Energy Agency - Recommended Practices, American Wind Energy Association and other national performance standards now estimate annual energy output for a turbine at several wind speeds, considering variations in the velocity to predict the performance. The projected energy production is a better measure of effectiveness than peak power rating.

Systems are generally designed to operate between wind speeds of 4 m/s and 25-30 m/s. The lower speed is the design cut-in when the turbine may begin to operate. The higher range is the design point where operation will not damage structural and subsystem components. Some turbines that employ stall control can operate without having to shutdown at excessive wind speeds.

Classification of wind turbines varies from country to country. For this report, the following classification applies:

i. Mini-systems - output less than 10 kW (electric or equivalent);

ii. Small systems - 10 to 100 kW;

iii. Medium systems - 100 to 500 kW;

iv. Large systems - Over 500 kW.

The following presents key characteristics of turbines installed during 1984 by several companies developing windfarms in the United States.

—   40 kilowatt turbine rated at 13.4 m/s wind speed. Cut-in speed 3.6 m/s, cut-out 22.3 m/s. Rotor diameter equals 13.4 m. Downwind turbine, three blades, free yaw, blade tip brakes and self-supporting tower;

—   95 kilowatt turbine rated at 17.9 m/s wind speed. Cut-in speed 5.4 m/s. Rotor diameter equals 11.0 m. Downwind turbine, three blades with blade tip brakes and guyed pipe tower;

—   100 kilowatt turbine rated at 12.9 m/s wind speed. Cut-in speed 5.4 m/s, cut-out 18.3 m/s. Rotor diameter equals 17.1 m. Down- wind turbine, with free yaw, variable pitch, computer control and tripod tower;

—   250 kilowatt turbine rated at 17.0 m/s wind speed. Cut-in speed 6.3 m/s, cut-out 26.8 m/s. Vertical axis machine with rotor diameter of 18.3 m.

The Danish systems that entered the United States windfarm market were three blade machines with upwind rotor, and induction generator. Danish manufacturers generally prefer active yawing systems for control and protect against overspeed by employing stall control with fixed pitch blades. These machines usually use a rigid rotor, cast hubs and forged shafts. Nacelles and towers are usually welded steel.

A limited number of experimental wind machines rated at one megawatt or larger have been built. As of 1985, eleven such systems existed in the world: seven in the United States, two in Sweden, one each in Denmark and Germany. Canada, Denmark, the Netherlands and the United Kingdom have plans for or also were constructing such turbines. It is premature to describe a standard machine in this size, except to note that all constructed to date have been horizontal axis units. One exception will be a new vertical axis 4 MW_e turbine being built in Canada.

III.   APPLICATIONS

In the past, the primary output was mechanical power, used for pumping water, for food processing and in some cases, for desalination systems. Wind systems used for these applications would generally be multi-bladed turbines which operate on drag forces, thereby achieving high starting torque and the capability of delivering energy even at low wind speeds.

The mechanical energy of a wind machine may also be used to produce heat directly or make electricity for heating homes or commercial buildings. Such heat-producing systems may also be used for agricultural applications: crop drying, processing foods, and heating for greenhouses. Heating technology applications have not advanced, however, because the electricity produced by such systems is more valuable for other uses than conversion to heat.

The primary wind energy application responsible for increased R&D over the last decade is electricity production. Applications for electricity from wind are mostly a function of the rated capacity of the turbine.

— Mini systems (1-2 kW or less) are often used in remote power applications. Frequently used as battery chargers, the systems meet the needs of small communication systems, navigation or military equipment, or even supply power for small businesses or homes.

— Residential/agricultural systems (3-40 kW) satisfy building energy needs and commercial requirements or provide energy for grain drying, water pumping and process heat. They can also be part of a power system that includes a diesel generator, or a photovoltaic system.

— Utility scale systems (15-1000 kW and above) generate electricity, either as stand-alone units in remote locations or grouped and interconnected in large numbers to feed an electricity grid, in which case they are called windfarms. Figure 23 shows a windfarm installation in the Netherlands.

IV.   TECHNICAL STATUS AND ACCOMPLISHMENTS

Status in 1974

In 1974, most installed wind machines were simple mechanical water pumpers. A few electricity producing machines were still operating, but serious wind manufacturers had already disappeared by the 1960s. There was a small base of scientific and engineering knowledge gained from some large turbine experiments conducted in Europe and the United States in the 1940s and 1950s. And the 1930s technology, once commercialized, could have been easily duplicated by modern industrial firms, had there been interest. Very little attention was paid to this technology, however, and the pool of trained engineers and aerodynamic specialists in wind energy was limited.

Technical Advances

When the oil crisis sharply raised conventional energy prices in 1974, many countries initiated wind energy research programmes. Many of the new entrants in the modern era of wind energy development came from aerospace corporations, who applied their knowledge of aerodynamics and structural engineering to wind turbines. Their entry was accompanied by some old-line industrial manufacturing firms, and quite a few entrepreneu-

FIGURE 23

REALISATION OF "WIND FARM" CONCEPT, CONSISTING OF EIGHTEEN WIND TURBINES EACH OF 300 kW. THE SITE IS NEAR TO SEXBIERUM FRIESLAND, NETHERLANDS

rial engineers who started new companies, some of which have grown to become industry leaders. Over time, the combination of extensive government research programmes in some countries and industrial interest developed a new generation of wind technology specialists. As the decade progressed, utilities also began to support research and consider the applicability of wind turbines to their power generation needs. During the decade following the oil crises, considerable infrastructure developed, leading to the installation of well over 10 000 wind machines worldwide by the mid-1980s. Half of these turbines used rotors greater than 13 meters in diameter, making them larger than all but a few experimental machines built prior to the mid-1970s.

In order to expand gains from national and industrial research programmes, the International Energy Agency developed two agreements in the wind energy field. One concerns generic wind energy conversion systems research and development, the second addresses large-scale system development. As a result of national R&D programmes and international co-operation, wind energy research has advanced along a broad front, logging major accomplishments in many areas. In addition, private industry-sponsored development efforts have led to commercialization of wind energy equipment in many countries. The initiatives begun in the 1970s have created wind technology manufacturing industries in several countries that now deliver systems that have gone through several generations of development.

A large number of technical achievements have been logged over the last decade. These have occurred in aerodynamics, resource measurement and siting, structural design, stress analysis and engineering, and in material sciences. Important progress has been made in the power conversion (interconnection) and electronic control fields as well. As a result, wind technology is fairly mature for the mini and small systems.

As turbines grew in scale to increase their efficiency, need for technical gains became apparent in three areas: regulation of the power output of the turbine; stress and fatigue and system lifetimes; and analysis of load dynamics. Governments and industry built and tested many different wind machines. Failures occurred, especially with the very largest turbines, where new boundaries in physical and aerodynamic sciences were being established. Nonetheless, the result has been development of medium-scale machines that have logged availabilities of 95% to 98% and capacity factors near 30% in ideal wind sites.

Among the innovations that have increased the economic and technical potential of wind energy, one key development was the synchronous inverter. It helped to open the way for small electric power systems to be interconnected to utility grids. The induction generator, now used widely in windfarm units, was another major adaptation that helped make smaller wind systems simpler and less expensive. Development of fibreglass-reinforced polyester (FRP) as wood/epoxy blades also led to rapid advances since series production of such aerodynamically sound blades allowed industrial firms unfamiliar with wind dynamics to become involved in the industry. A further advance in fibreglass winding techniques now makes it possible to produce very large blades with automatic machinery in less time. In addition, some firms have turned to large laminated wooden blades, which have increased performance and lifetimes of wind turbines. Finally, it was discovered that stresses on wind machine airfoils were much greater than previously experienced in practical aerodynamics and that micro variations existing in wind streams profoundly affect rotor structural loads and performance and integrity. The teetered rotor and flexible towers and drive trains were still other important innovations which have helped solve the structural load problems large rotors experience.

Increases in energy production of wind turbines installed commercially have demonstrated the growing maturity of wind energy technology. In Denmark, average energy production in 1985 from field installed 55 kW turbines was near 100 MWh/year, double the 1980 figure (50 MWh). This gain is due to improved design and production techniques, leading to greater reliability; taller towers; an increase of 5% in swept area; increases in aerodynamic and transmission efficiencies; and better site selection. Commercially ready 75 kW and 95 kW turbines are demonstrating new gains, producing from 600-700 kWh/m²/year compared to 560 kWh for the latest 55 kW machines.

The Danish technical programme concentrated on two areas: first, to overcome fatigue problems encountered in wind turbines and second, to reduce installation costs. Results have led to highly reliable, cost-competitive small- and medium-sized wind turbines for windfarm and stand-alone applications. With the Danish national programmes and support, Danish industry has achieved a leading position in the commercialization of wind energy, and developed a worldwide reputation for sturdy, robust turbines.

In the United States, performance monitoring of windfarms has shown dramatic improvements in technology and performance. The average size of turbines being installed increased from 49 kW in 1981 to 78 kW in 1984. Over the same period, availability increased from 0%-70% to 80%-95% for various windfarms. More mature, well-maintained systems are achieving availabilities of 95%-98%. Capacity factors in windfarms have also risen. Pacific Gas & Electric Company's monitoring of windfarms in the Altamont Pass has revealed the gains: 0.03 in 1982, 0.05 in 1983, and 0.12 in 1984. Performance for 1985 ranged from 0.13 to 0.25, with an average of 0.19. Lower capacity factors in the first years of windfarm development were due in large part to completely inoperable equipment installed in the fields. Over time, such turbines have been removed or refurbished, increasing performance of the windfarms. Projections are that capacity factors of 0.30 to 0.35 are possible at good wind sites and even higher at sites with annual wind speed average in the 10 m/s range.

Development of large turbines over one megawatt has been slower, more expensive and more difficult than that of smaller machines. Tests and demonstrations of large turbines have been marred by equipment malfunctions, material failures and delays in operational data collection required for further development. Performance of experimental large systems has shown that they were much more complex to develop than anticipated. Aerodynamic performance and created stresses are greater than those experienced by aircraft. A MOD-2 purchased by Pacific Gas & Electric Company in California has required modifications or repair of the yaw drive, tower, low speed shaft, various sensors, shaft bearing seals and lubrication systems. The pitch control system required modification to remain stable at rated power under all wind conditions. After repairs and fixes this turbine has logged a capacity factor of 0.16, and tests indicate this figure could reach 0.29 to 0.35 annually. The WTS-3, 3 MW turbine in Sweden has successfully operated more than 9 500 hours and produced 14 million kWh.

Denmark has designed and installed two different 40 m rotor (rated 630 kW) wind turbines, Nibe A and Nibe B, commissioned in 1979-80. A key finding from this programme was that blade surface irregularities on a large-scale wind turbine have a significant impact on performance. The Nibe experiments have convinced Denmark that large turbines can be built and operated safely and reliably and with greater economy than small turbines, but that more testing of several generations is necessary before technical maturity is realized. Meanwhile the first commercialization steps for large turbines are underway. Several utilities (e.g. ELSAM, ELKRAFT, FYNS-VAERKET) have commissioned such turbines, funded partly by the power companies, the Danish Government and the Commission of the European Communities. Five 750 kW turbines will be installed in one windfarm.

Other country-specific advances which may have significant impact include Germany's major experimental project, GROWIAN, a 3 MW turbine with a 100 m rotor diameter (the world's largest). This unit has been under test operation since the summer of 1983. Although cracks developed in highly stressed parts of the hub that required repairs during the testing period, German scientists feel the results have confirmed that many of the technical risks of such a large machine are controllable. The United States is developing an advanced multi-megawatt machine called the MOD-5. It will have a 3.2 MW variable speed generator and a 98 m rotor diameter. This machine will begin operating late in 1987.

In the field of large wind turbines, the Spanish-German project AWEC-60, for developing of a 3 blade, 60 m diameter horizontal axis wind turbine of a rate power of 1.2 MW at 12.2m/s is being carried out, sponsored by the CEC. The American and European turbine installations over 1 MW_e in size have begun to accumulate significant hours of operation, and all have provided valuable experience. Still, despite the progress, the pace of development of such large turbines has clearly slowed.

Commercial Status

Wind energy is not yet competitive with conventional baseload electricity generating technology. However, with incentives, it is developing commercially in a few markets where adequate wind resources coexist with high marginal energy costs for satisfying peak demand and/or where there is an interest in supply diversification. Wind energy is competitive for small remote power systems in suitable locations. Large utility scale turbines (more than 1 MW) are still in an advanced stage of R&D.

The combination of high energy costs, over reliance on oil and gas, wind regimes with average velocities of 7 m/s, modifications in utility regulations, and federal and state incentives for investors created the largest wind energy market in the world in California, United States. This commercial development is overwhelmingly in the area of multi-turbine installations in windfarms. A similar set of circumstances has begun to create markets in other countries also, most notably Denmark. Table 19 tabulates where wind-generated electricity was produced in the world in 1985.

TABLE 19

WIND ENERGY PRODUCED IN 1985 BY LOCATION

CALIFORNIA INVESTOR OWNED UTILITIES[1]	
- Pacific Gas & Electric Co.	50.0 %
- Southern California Edison Co.	38.0 %
DENMARK	5.3 %
CALIFORNIA DEPT. OF WATER RESOUR.	1.6 %
OTHER EUROPEAN COUNTRIES	1.3 %
OTHER UNITED STATES LOCATIONS	1.3 %
REST OF THE WORLD	2.8 %

1. Interconnected to, but not owned by three utilities.

Source: American Wind Energy Association

1985 Market Development At the end of 1985, 1 120 megawatts of wind capacity had been installed in the United States, with nearly 99% of it in California windfarms. Wind energy produced in 1985 amounted to 662 million kilowatt hours. The American Wind Energy Association reports a saving equivalent to 1.1 million barrels of oil from the 13 000 wind machines. In California, wind installations represent 2.5% of installed electricity generation capacity and are meeting nearly 1% of demand.

As the windfarm market in the United States has developed, manufacturers have steadily increased the power rating of the equipment. In California, in 1981, the average size turbine was rated at 52 kilowatts of capacity. By 1985, the average had increased to 112 kilowatts, pushed upward by the introduction of 75 turbines from a European manufacturer, each rated at 330 kilowatts of capacity. A United States manufacturer also installed turbines in Hawaii, United States, rated at 600 kilowatts.

When expressed in terms of rotor swept area, installations in the three largest windfarms in California show that United States manufacturers held slightly more than 50% of the market. Danish equipment accounted for 41%.

Belgian, German, Irish, Netherlands and Scottish companies accounted for the remaining 7%. Table 20 shows the market share of leading manufacturers. Of these companies, only Flowind manufactures vertical-axis turbines.

In Denmark, market development began with individually-installed, grid-connected wind turbines, and then windfarm development began a modest take-off. While approximately 1 400 individual units have been sold, the last two years have also seen private individuals, municipalities and public institutions build nine windfarms, and four more are under construction. Turbines in the windfarms range from 55 kW to 95 kW rated capacity. Five 750 kW and three 300 kW turbines are being installed in two of the four new fields under construction. Thus, commercialization of large turbines is beginning. Annual electricity production from the 116 turbines (7.6 MW of capacity) in the existing windfarms is expected to total 12 700 000 kWh/year. The market development has been fairly rapid, as installed wind capacity has risen from 5 MW in 1980 to 36 MW in 1984, then nearly doubled to 62 MW in 1985. This wind power, for which the government provides one third to one half of the cost, is providing from 0.3% to 0.5% of electricity used in Denmark.

TABLE 20
**LEADING MANUFACTURERS OF
COMMERCIALIZED WIND TURBINES**
(Based on cumulative installed rotor swept area, 1985)

Company	Percent of Market
United States Windpower (US)	20
Vestas (DK)	13
Micon (DK)	12
Flowind (US) (vertical axis)	7
ESI (US)	7
Fayette (US)	6
Bonus (DK)	5
Nordtank (DK)	4
Wind-Matic (DK)	4
Enertech (US)	3
Howden (UK)	3
Others	18

Source: American Wind Energy Association

Risø National Laboratory in Denmark reports that the cost of a 75 kW turbine is as shown in Table 21.

Assuming maintenance cost of US $1 350 per year, 20-year life and an annual production of 150 000 kWh, the estimated energy costs for this turbine would equal US $0.0625/kWh.

There is growing interest in windfarms in other countries as well, frequently spurred by government initiatives and utility companies. A Belgian manufacturer has concluded an agreement for a 5 MW windfarm on an off-shore dike with twenty-five 200 kW turbines. Companies in the Netherlands are proposing a 5 MW windfarm, and a German firm has installed a windfarm in Greece to help develop a Mediterranean market.

TABLE 21
DANISH 75 kW TURBINE INSTALLED COSTS
(US $1985)

Wind turbine price	58 125
Foundation	6 500
Installation	5 000
Grid connection	5 000
Consultancy	1 000
Miscellaneous	1 250
Total	US $81 875

Standards and Certification In several countries, industry and government have co-operated to institute testing and certification standards for wind energy equipment. In the United Kingdom, a National Test Centre for small and medium wind turbines was established by the National Engineering Laboratory in Scotland. In Denmark, the Risø test centre implemented strict standards for commercial turbines. In the United States, the California Energy Commission has established a windfarm energy performance reporting programme, and the federal government has supported the American Wind Energy Association in development of industry voluntary standards. The IEA has supported the development of recommended practices in seven areas including: power performance testing, cost of energy computation, fatigue evaluation, acoustics, electromagnetic interference, safety and reliability and quality of power. These and other programmes mark an important step in the commercialization of wind energy technology.

V.   ENVIRONMENTAL CONSIDERATIONS

The environmental impacts associated with individual wind machine installations are not significant and are certainly restricted to the local area. Although effects of large windfarm installations are more pronounced, they are not characterized as environmentally threatening. These effects, for both individual and windfarm machines may be summarized as follows:

—   Safety hazards from operations and/or failures of rotating systems;

—   Electromagnetic interference produced by rotating structures;

—   Visual or aesthetic impact;

—   Acoustic noise;

—   Hazard to birds;

—   Land erosion due to construction and maintenance traffic;

—   Hazard to endangered species due to construction and maintenance.

Very large wind turbines could pose some hazard to aircraft, but the risk of a serious accident is quite small. Extreme weather conditions or fatique causing rotors to disintegrate and throw blades is more of a concern, but such problems are not considered a serious constraint for the technology. Wind turbines will also throw ice long distances. These problems are controllable through implementation of standards governing safe design and operation.

Interference with radio and television signals is a concern for inhabitants in the vicinity of installations with medium and large wind machines. While all forms of electromagnetic communications may be affected, signals from the more directional systems (e.g. microwave) that are not in line of sight should be immune to such interference. These affects are local in nature, usually not extending for more than 2-3 km from the largest turbines. Measures to handle the problems include siting wind machines away from areas where they will interfere, moving communication equipment or installing slave transmitters or cable networks.

Visual or aesthetic problems have increased as the number of windfarms has expanded. The largest such installations in California number thousands of turbines, and local residents have objected to the resultant changes in their visual environment. The impact of single machines is not likely to be as pronounced.

Noise from wind turbines can disturb local residents. The largest multi-megawatt turbines can be audible up to 2 100 meters downwind. Noise levels of 56 dB(A) have been recorded at distances up to 400 meters. Gear box and generator noise is understood and controllable. Noise from the fluctuating lift forces on turbine blades, especially from downwind turbines in the shadow of towers, is an area where more research is required. Siting of wind machines may require some of the same considerations given to noisy industrial facilities.

Rotating systems may pose a hazard to birds, but this concern does not appear likely to present a constraint to the use of wind energy.

The largest environmental impact from wind energy is likely to come from the construction of large windfarms, but the effects will be limited to the local area. The adverse impacts on the land may be dealt with through regulation. For example, local jurisdictions in California require windfarm developers to restore sites once equipment has been installed. Once erected, it is not expected that large wind installations will have any significant ecological impact on nearby areas due to the micro climatic changes they may cause in wind patterns. With regard to materials, wind energy technology poses no unusual environmental threats since such systems are constructed of commonly used industrial goods. Disposal of retired systems should cause no additional requirements for the secondary materials markets that already exist.

In summary, the environmental impacts associated with wind energy technology are highly local in nature, and are not considered to be serious constraints to widescale use. Visual and noise pollution and television interference have surfaced as the issues of greatest concern. Public risk from accidents is most likely to occur with small dispersed systems, not large windfarm installations. Careful siting, sound machine design, and attentive construction and operation practices should mitigate or eliminate many of the negative environmental and safety effects.

VI.   TECHNOLOGY OUTLOOK

Technological Prospects

No insurmountable technical barriers appear likely to block further improvement of wind energy technologies. Despite the progress already evident, research needs remain quite diverse, and few areas can be described as completely mature. Areas that should benefit from continued research are aerodynamic prediction codes for turbulent conditions and studies on dynamic stall and thick airfoils. Blades can benefit from new research on materials and structural dynamics and more information on mechanical fatigue should also assist industry to develop longer-lived products.

Intermediate turbines suitable for windfarms, and large, multi-megawatt systems are now receiving the greatest level of R&D funding and should advance most rapidly. The commercial market for small systems does not appear strong enough to support the extensive industry R&D required to increase their competitiveness. Sales for any manufacturer often number in the hundreds, and until large-scale production and technological innovation can reduce costs, the technical potential of small systems could lag. They should nevertheless benefit from technical advances made in basic wind research.

Wind economics depend most on the yearly output of a turbine, the investment cost and the annual operating and maintenance costs. The prospects for improving wind energy systems in these three categories appears favorable.

Yearly output is a function of efficiency and reliability. Additional efficiency improvements should be forthcoming as governments and industry are still supporting scientific research and engineering to advance airfoils, controls and drive trains. Efficiency will also increase as turbine sizes increase, a trend that gives no indication of slowing. In addition, basic research on atmospheric fluid dynamics, aerodynamics, and structural dynamics promise to increase the capability and sophistication of design and analysis tools. Finally, several governments continue to fund research in advanced system concepts.

Improvement of existing technology appears likely because most turbines meet their claimed power curves at low wind speeds and because manufacturers have straightforward means to improve and reach full power at the rated wind speed. Analysts observe that improvement can come from alteration of blade pitch angles, use of different rotor speeds and airfoil profiles or vortex generators on blades, and fine tuning of control strategies to eliminate losses due to gusting winds shutting down turbines when they could still be operating. None of these tasks are formidable in nature, and Danish engineers have observed that an improvement in output of 15% from current technology (75 kW system) combined with a 1.7% annual real price rise for coal should make wind energy competitive with fossil fuel electric power plants in Denmark by the year 2005.

The United States wind energy programme five-year development plan has fixed the current status of wind technology and established technical goals in five key areas, summarized in Table 22.

TABLE 22
TECHNOLOGY STATUS AND TECHNICAL GOALS
(US $1984)

Technology Characteristics	Current Technology	Technical Goals
Annual energy production (kWh) and installed costs (dollars) per square meter of swept rotor area	Ranging from: 300 kWh/m² at US $600/m² to 800 kWh/m² at $1100/m²	Ranging from: 900 kWh/m² at $450/m² to 1500 kWh/m² at $750/m²
Annual operation and maintenance costs	2¢/kWh with	0.5¢/kWh with
Availability	85%-95%	95%+
Life-expectancy . Major fatigue-related parts . Balance of the primary structure	5 years 30 years	30 years 30 years
30-year levelized cost of electricity	10¢/kWh-15¢/kWh	4¢/kWh

NOTE: Current technology represents state-of-the art machines: i.e., machines with the best combination of all characteristics leading to lowest electricity cost, and not necessarily the best machines measured by any one characteristic such as highest energy production. Energy production for current machines was estimated for an annual average wind speed of 6.3 m/s (14 mph) at 10 m above ground to allow comparison with the long-term goals. Electricity cost estimates exclude solar tax credits.

Outlook

Given its present status and the remaining opportunities to expand technical potential of wind energy, the outlook for continued progress is favorable. While wind should be able to compete in the near term in the existing market for high-priced power, its future lies in competing in a wide variety of applications. It could then become a component of a diversified energy supply for many countries.

Utilities have demonstrated a significant level of interest in wind energy technology, but they have also expressed concern about the stochastic nature of the resource. California utilities find that windfarm production occurs during peak demand periods, but also extends beyond these hours, when cheaper sources of energy may be available. Utilities in Denmark are concerned about interconnection strategies being able to cope with power surges due to varied wind conditions. Some utilities have estimated that no more than 10% of their power needs could be met with wind technology because of its intermittant patterns.

A key development affecting near-term market development has been the expiration of federal government tax credits for wind energy systems in the United States. Industry analysts predict sales in 1986 could drop to the 1984 level, and Danish exports in 1986 are reported to be 50% of the 1985 figure. A serious contraction of the existing wind energy industry in the United States appears to be underway. Severe competition for market share will lower prices, depriving companies of income required to further improve technology. The overall effect of the American market slowdown could be a deceleration of technology development and progress. This development, however, could spur the opening of new markets on the Atlantic coast, Mediterrean regions in Europe, and in developing countries.

Expiration of tax credits for residential owners of wind machines will further depress the declining market for small wind machines in the United States. The market for small systems in the United States, especially for residential applications had already dropped signficantly in 1985 compared to the early 1980s. Initial capital costs for equipment must be further reduced for this market to expand in other than remote sites with excellent wind speeds.

In Europe, there appears to be increased interest in developing windfarm projects similar to those in the United States in order to assess the longer range future of wind power. Wind farms are being installed in several countries including Denmark, the Netherlands, Great Britain, Greece and Spain. In Denmark, Government and utilities have agreed that 100 MW of new wind energy systems should be installed between 1986 and 1991. One company will install five 750 kW turbines (40 m rotor) in 1986, while another plans to install a 2 MW unit (60 m rotor). The Government's goal is for renewable energy sources - with an emphasis on wind turbines - to provide 8% of Danish fuel consumption in the year 2000. The target for wind in 1995 is to satisfy 10% (4 TWh) of national electricity consumption. In Spain, four small wind farms, with Spanish wind turbines, will be installed in 1986 and six more in 1987.

In Germany, the government has continued support for wind energy research and German companies have been slowly developing export markets. Recent studies on the economic and exploitable potential of wind energy for the year 2000 arrived at an economic potential equal to 12% of German electricity consumption in 1980 and an exploitable potential of 1.3%-1.8%. This penetration represents about 500 large-scale wind energy systems in the MW class (see Table 23).

TABLE 23

WIND ENERGY POTENTIAL IN THE YEAR 2000 IN GERMANY

Relative to 300 TWh/a (approx. 1980 FRG elec. consumption)

Technology	Potential		(No. of Plants)
	Economic	Exploitable	
Case I			
- Large Turbines	12.2 %	1.3 %	(345)
- Small Turbines	0.17%	0.02%	(7800)
Case II			
- Large Turbines	12.2 %	1.8 %	(500)
- Small Turbines	0.45%	0.05%	(34500)

Case I: Elec. Price: +2%/a; "high" interest rate
Case II: Elec. Price: +3.5%/a; "low" interest rate
Large: = 1 - 3 MW. Small = 20 - 100 kw

Source: Study commissioned by the Federal Minister of Economics and carried out by the Deutsches Institut für Wirtschaftsforschung (DIW) in Berlin and the Fraunhofer-Institut für Systemtechnik und Innovationsforschung (ISI) in Karlsruhe, Bonn, 1985

In summary, when good wind resources exist in remote sites, cost effective applications of current wind energy technology may be realized. Grid-connected power applications have grown rapidly. Initial markets that are developing in northern European countries and in the United States are expected to expand into the 1990s. The rate of growth experienced in the United States market - which has enabled many European firms to develop commercially - may slow, however, through the balance of the 1980s. Markets should expand in developing countries where fossil energy is utilized for electricity generation. Ultimate potential depends on future improvements in wind power cost and performance, and price movement of conventional energy sources.

VII. MAJOR FINDINGS

Significant advancements have been made in wind energy conversion technology over the last decade. Wind systems are currently operating successfully in a large number of countries. A high degree of progress has been witnessed in reliability and availability gains for commercially installed equipment. The market growth that has occurred has been due to a combination of factors: location of excellent wind resources in high cost energy areas; over dependence on oil and gas; government policy designed to encourage alternative energy use in the utility sector; and government investment incentives attracting capital to large projects. Utility interconnected projects (windfarms) have been the dominant application.

Significant reductions in equipment and installation costs due to machine design improvements, production volume increase and competition which resulted as windfarm development expanded have continued in the mid-1980s. From 1981 to 1985, certain costs have been reduced as much as 40% in constant dollars. The cost of delivered power has dropped by a factor of ten over the recent development period.

Companies have shifted to production of larger turbines to achieve certain economies and to meet demand for windfarm systems. The up-scaling has contributed importantly to the reduction in costs. The highest growth rate exists for turbines over 100 kW in size, and by 1985 such turbines accounted for more capacity being installed than any other size.

Sales of small turbines have lagged far behind the windfarm market, even declining in some countries. Manufacturers expect these sales to recover because electricity prices continue to rise in many countries.

The wind market has been highly concentrated with nearly 90% of wind energy produced in the United States. Manufacturers, however, are not nearly as concentrated with many companies competing for the market.

Environmental effects have been identified that require additional research and development to make wind systems more acceptable. Environmental concerns, though, are not considered a barrier to commercial expansion of wind energy. Standards, certification procedures and performance reporting systems have been put in place in a few IEA countries.

Pay-offs from current research plans and development should further improve wind capture potential, reliability, life expectancy, and availability. Positive outlook towards future wind energy potential has many countries continuing to perform wind energy resource assessments, R&D and demonstration projects. Improved modelling techniques should allow more accurate assessments of wind energy potential in the coming years.

VIII. BIBLIOGRAPHY

1. *Achievements of the European Community Second Energy R&D Programme* by J.T. McMillian and A.S. Strub, EUR 9204 EN, FR 1984.

2. Annual Report 1985, *IEA Implementing Agreement for Co-operation in the Development of Large Scale Wind Energy Conversion Systems* (LS WECS).

3. Annual Report 1985, *IEA Implementing Agreement for a Programme of Research and Development on Wind Energy Conversion Systems* (WECS).

4. *Recommended Practices for Wind Turbine Testing and Evaluation:*
   Part 1: Power Performance Testing
   Part 2: Cost of Energy from WECS
   Part 3: Fatigue Evaluation
   Part 4: Acoustics
   Part 5: Electromagnetic Interference
   Part 6: Safety and Reliability
   Part 7: Quality of Power

   submitted to the Executive Committee of the International Energy Agency Programme for Research and Development on Wind Energy Conversion Systems.

5. Chappel, M.S., *Status and Outlook for Wind Energy Research and Development in Canada* CIRCA 1984, NRCC Dept. No. 24202 (Francais CNRC Report No. 24687), December 1984.

6. The Test Station for Windmills, Publication from Risø National Laboratory, P.O.B. 49, DK-4000 Roskilde.

7. *The International Energy Agency Wind Energy Co-operation,* B. Pershagen, Studsvik Energiteknik AB, S-61182 Nyköping, Sweden European Wind Energy Conference, October 23-24, 1984, Hamburg.

8. *Wind Energy Evaluation for the CEC,* P. Musgrove, Reading, November 1982, CEC Ref No. XVII/AR/82/255.

9. Jahresbericht 1985 Uber Rationelle Energieverwendung, Fossile Energieträger, Neue Energiequellen (Annual Report 1985 of the Programme for Energy Research and Technologies), Projekleitung Energieforschung (PLE), KFA Jülich. Verlag Fachinformationszentrum Energie, Physik Mathematik GmbH, Karlsruhe.

10. Wind Power: *Results and Conclusions from the Swedish Wind Energy Programme,* National Energy Administration 1985:1 (in Swedish with an English Summary).

11. *Wind Energy, Five Years R&D Plan,* United States Department Of Energy, January, 1985.

12. *Solar Energy Industries Association, Energy Innovation: Development and Status of the Renewable Energy Industries 1985,* Volume II, Washington, 1986.

13. Paul Gipe, *Wind Energy: How to Use It,* Stackpole Books, Harrisburg, PA, 1983.

14. American Wind Energy Association, *Proceedings: Wind Energy Expo 1983 and National Conference, October 17-19, 1983,* Alexandria, VA, 1983.

CHAPTER FOUR: BIOMASS ENERGY

A. NATURE OF THE BIOMASS ENERGY SOURCE

Biomass is all the matter that can be derived directly or indirectly from plant photosynthesis. "Indirectly" refers to the products available via animal husbandry and the food industry. The resource base includes hundreds of thousands of plant species, terrestrial and aquatic, various agricultural and industrial residues and process waste, sewage and animal waste. Municipal waste may be considered biomass, but is largely excluded in this review. The essential resource is carbon fixed on a renewable basis through the photosynthetic process. The resource is vast, 170 billion tons of dry biomass are produced world-wide each year, whereas global consumption is currently 1.3 billion tons. The potential impact of biomass resource derived energy (bioenergy) on global energy needs is large.

Biomass energy conversion is perhaps the most technically, economically and socially complex renewable energy option. The resource base is highly diverse in terms of its availability and its physical and chemical properties. Generally bulky and expensive to transport, biomass has an economically limited collection radius that places limits on where it can be used. Realization of its potential involves resource management (i.e. land, water and nutrients); technology development and technology transfer; and awareness of and possible adjustments to the social and political environments.

A fundamental distinction in the use of biomass exists between the conversion of waste and residue, which increases the energy balance and efficiency of existing economic activity involving biomass (e.g. paper industry, farming), and the cultivation of biomass for energy, which then requires all the inputs of an agricultural production system: land, labor, equipment, nutrients, etc. As exploitation of this latter area has expanded, it has raised agricultural policy issues regarding such concerns as support for conventional agricultural products versus support for energy feedstock production conversion. In short, politics can have a profound effect on the availability of biomass resources.

Biomass resources used for energy may be divided into three categories: forestry and wood processing residues, crop residues and animal wastes, and energy crops. Agricultural and forest residues are the sources of biomass most available for conversion to energy. Peat moss is harvested in a few countries for heating and electric power generation.

Wood Resources

Wood is considered the biomass resource with the greatest near-term potential in Western Europe and North America. Land mass, population, and productive forest area of the IEA Member nations are summarized in Table 24.

TABLE 24

POPULATION, LAND AND FOREST AREAS IN IEA COUNTRIES

Country	Land Area 1000 km²	Population		Productive Forest Area	
		Total Mill	Per km²	Total Mill. ha	ha per capita
Australia	7 686	15.7	2	40.8	2.6
Austria	84	7.6	90	2.95	0.39
Belgium	31	9.9	319	0.62	0.06
Canada	9 900	25.0	3	200.00	8.00
Denmark	43	5.1	119	0.41	0.08
Finland*	337	4.7	14	19.70	4.19
Germany	248	62.2	251		
Greece	132	9.2	70		
Ireland	70	3.4	49	0.35	0.10
Italy	301	56.5		188	
Japan	377	114.2	303		
Luxembourg	2.6	0.36	138		
Netherlands	41	13.9	339		
New Zealand	269	3.2	12	1.00	0.31
Norway	326	4.1	13	6.66	1.62
Portugal	92	9.8	107	2.76	0.27
Spain	508	38.8	76	19.8	0.51
Sweden	412	8.3	20	23.40	2.82
Switzerland	41	6.6	161	1.18	0.17
Turkey	779	42.0	54		
United Kingdom	244	56.1	230	2.2	0.04
United States	9 125	230.0	25	195.00	0.85
Total/Average	31 049	724.3	23		

* Finland participates in IEA Forestry Energy Programme

Source: Time Almanac 1980 or data provided by Member countries

The ultimate energy potential of wood must take into consideration its value as a replacement for more energy-intensive materials such as plastic or metal. Assessing wood resources, therefore, involves determining the economic value for other purposes and the accessibility and conversion potential of the residues. The IEA Expert Group on Renewable Energy Policies gathered data presented in Table 25 to estimate the energy potential of the forestry and timber industries. These data indicate a very large difference in energy potential from country to country.

In several IEA countries, wood-related industries already meet 30% to 50% of their energy requirements through conversion of wood residue. For example, Finland has scarcely any residues unused. On average, about 25%

TABLE 25
AVAILABLE ENERGY POTENTIAL FROM FORESTRY
AND TIMBER INDUSTRY RESIDUES (1979)[1]

Country	Timber Production $10^6 m^3$	Residues[2] (Mt)			Energy Potential (Mtoe)		
		Forest	Industry	Total	Forest	Industry[3]	Total
Canada	160.6	53.5	13.3	66.8	25.7	6.4	32.1
United States	342.3	114.0	28.7	142.7	54.7	13.8	68.5
Japan	34.0	11.3	2.8	14.1	5.4	1.3	6.7
Australia	15.5	5.2	1.3	6.5	2.5	0.6	3.1
New Zealand	9.0	3.0	0.8	3.8	1.4	0.4	1.8
Austria	14.0	4.7	1.2	5.9	2.2	0.6	2.8
Belgium-Luxembourg	2.7	0.9	0.2	1.1	0.4	0.1	0.5
Denmark	2.0	0.7	0.2	0.9	0.3	0.1	0.4
Finland	43.1[3]	14.4	3.6	18.0	6.9	1.7	8.6
France	30.7[3]	10.2	2.6	12.8	4.9	1.2	6.1
Germany	31.4	10.5	2.6	13.1	5.0	1.2	6.2
Greece	2.9	1.0	0.2	1.2	0.5	0.1	0.6
Ireland	0.3	0.1	-	0.1	-	-	-
Italy	7.7	2.6	0.6	3.2	1.2	0.3	1.5
Netherlands	0.9	0.3	0.1	0.4	0.2	-	0.2
Norway	8.5	2.8	0.7	3.5	1.3	0.3	1.6
Portugal	8.6[3]	2.9	0.7	3.6	1.4	0.3	1.6
Spain	11.9	4.0	1.0	5.0	1.9	0.5	2.4
Sweden	51.7	17.2	4.3	21.5	8.3	2.1	10.4
Switzerland	4.3	1.4	0.4	1.8	0.7	0.2	0.9
Turkey	29.1	9.7	2.4	12.1	4.7	1.2	5.9
United Kingdom	4.0	1.3	0.4	1.7	0.7	0.2	0.8

1. FAO Yearbook of Forest Products
2. Dry matter
3. Already partially used

Note: The values shown are calculated according to very simple basic assumptions, applied uniformly to each Member country and aim at giving simply a relative order of magnitude of the potential in the OECD countries. They may therefore differ slightly from estimates presented in other sources

Source: From Biomass Energy utilization opportunities. Biomass Overview paper for IEA Experts' Group on Renewable Energy Policies, prepared by Department of Energy, Mines and Resources and Trades (Canada), November 1981

of the wood entering the timber industry is available for conversion to energy.[1] Another main wood resource is fuelwood from farm woodland. This resource is used extensively for domestic heating in rural areas in many IEA countries. In sum, forest resources comprise the largest source of biomass feedstocks in both the developed and developing nations of the world.

Agricultural Residues and Wastes

Crop residues and animal wastes are a substantial by-product of the food-producing agricultural sector. Table 26 provides an estimate of recoverable energy potential from these resources based on assumptions about the ratio between principal products and by-products and the economics of energy conversion. The table is likely to overstate the energy potential since collection may be uneconomic and alternative, non-energy uses may be more valuable for the biomass in question. Using the most energy-efficient process, these resources could contribute from 0.4% to 6% of total energy requirements in IEA member countries, with the exception of Turkey, where the contribution could equal 22.5%.

Besides crop and animal wastes, another biomass resource is agricultural food surpluses, sometimes destroyed to support markets, and perishable and sub-standard products. It has been estimated that sugar surpluses in the European Community might replace 2% of fuel consumed if it were converted to ethanol. In the United States, corn surpluses might replace more than 7% of the United States gasoline consumption if it were converted to ethanol. Assessing the actual potential of agricultural surpluses and non-vendable produce as a biomass resource is difficult especially since diverse agricultural policies are also involved.

Energy Crops

A potentially more reliable resource would be energy-dedicated cultivation of biomass. Factors to be included in a resource assessment include the amount of available, both worked and unworked, arable land and available aquatic resources for growing plants traditionally used for food (beet, sugar cane, grains, artichoke, cassava) or non-food plants (bamboo reed, grasses, forest species) that possess great capacity for rapid growth, especially under marginal conditions. Such assessments have been made for many countries, but they are highly speculative. For example, if the output from all uncultivated land in Australia with potential to support energy plantations, together with the waste from existing agricultural and forestry production were used to produce methanol or ethanol, this alcohol could supply 60% of the transport fuel used (a theoretical maximum as this level of contribution is not attainable).

---

1. "Biomass Energy Utilization Opportunities", Biomass Overview Paper for IEA Experts. Department of Energy, Mines and Resources. Ottawa, Canada, November 1981.

TABLE 26

TOTAL AND RECOVERABLE POTENTIAL ENERGY PRODUCTION OF AGRICULTURAL RESIDUES AND WASTES 1978-79-80 Average in Mtoe

Country	Total Energy Potential			Recoverable Energy Potential			
	Crop residues	Animal wastes	Total	Crop residues	Animal wastes	Total	% of total energy requirements
Canada	15.9	2.1	18.0	2.3[1]	1.8[2]	4.1	1.9
United States	135.8	19.9	155.7	20.2[3]	5.0[3]	25.2	1.4
Japan	9.1	1.6	10.7	0.4[4]	1.2[5]	1.6	0.4
Australia	8.0	4.5	12.5	3.1[6]			
New Zealand	0.2	1.3	1.5				
Austria	1.3	0.5	1.8	0.4	0.3	0.7	2.8
Belgium-Luxembourg	0.8	0.7	1.5	0.1	0.3	0.4	0.9
Denmark	2.3	0.7	3.0	1.1	0.4	1.5	7.3
Finland	1.0	0.3	1.3	0.4	0.2	0.6	2.4
France	14.5	4.4	18.9	5.2	2.4	7.6	4.0
Germany	7.4	3.0	10.4	2.2	1.6	3.8	1.4
Greece	1.5	0.3	1.8	0.9	0.1	1.0	6.5
Ireland	0.6	0.9	1.5	0.0	0.5	0.5	8.2
Italy	6.6	1.9	8.5	2.5	0.9	3.4	2.5
Netherlands	0.7	1.2	1.9	0.1	0.6	0.7	1.1
Norway	0.3	0.2	0.5	0.0	0.1	0.1	0.5
Portugal	0.6	0.3	0.9	0.3	0.2	0.5	5.0
Spain	7.1	1.2	8.3	4.1	0.6	4.7	6.6
Sweden	1.6	0.4	2.0	0.8	0.2	1.0	2.0
Switzerland	0.3	0.4	0.7	0.1	0.2	0.3	1.2
Turkey	9.5	2.7	12.9	5.4	1.5	6.9	22.5
United Kingdom	5.4	2.5	7.9	1.3	1.3	2.6	1.2

1. Colwell, H.T.M., "The potential for a sustained energy supply from combustible crop residues in the Canadian Agriculture and Food system - A National Assessment".
2. Timber, G.E. Marshall, D., "Biogas as a farm energy source", Agriculture Canada, Engineering and Statistical Research Institute, Report I-297.
3. Tiner, W.E., "Biomass Energy Potential in the United States", Maginzira, Vol. 5/1
4. Available energy from grain chaff at present burnt or discarded.
5. Agriculture and Energy - Japan - Production of biogas from animal waste.
6. Potential methanol equivalent from grain and sugar cane residues, "Renewable Energy Resources in Australia", National Energy Advisory Committee, July 1981

Another example of the difficulty assessing such potential is found in analyses of energy crops in the EEC. Depending on the scenarios utilized, studies have estimated that between the years 1984 and 2000, anywhere from 36 Mtoe (metric tons of oil equivalent) to 105 Mtoe could be produced. The highest figure would require growth of cash crops on 2.86 Mha of land and use of 4.3 Mha of marginal land for energy crops; that is, 8% of total land area would have to be shifted to energy use. Exploitation of this biomass resource in this scenario would require not only application of the simple factors of production but also extensive political and social adjustment.

Aquatic plants offer another potentially large biomass resource. To date, they are used minimally for food or industrial products; however, micro-algae, macro-algae, floating and emergent plants could find extensive use as biomass feedstocks. For example, in the United States water hyacinth has been commercially produced at a rate of 11 dry tons per hectare per year and micro-algae growth rates equivalent to 20 dry tons per hectare per year have been demonstrated.

These energy biomass crops would be converted to gas or heat, liquid fuels, or energy intensive petrochemical substitutes. Resource potential also exists from unconventional biomass crops such as micro-algae species, from oil-convertible lipids (e.g. Euphorbia and milkweed) that directly produce hydrocarbons and from some desert crops (e.g. jojoba). The potential of these new approaches to expand the bioenergy resource base is uncertain but is being examined.

In summary, the diversity of biomass resources makes its ultimate energy potential somewhat difficult to assess at this time except to state that this resource will continue to be available to supply important quantities of energy to many IEA countries.

B. BIOMASS CONVERSION TECHNOLOGIES

I. INTRODUCTION

From pre-recorded history until the 19th century, humans relied on direct combustion of wood as their primary energy source. From the simple burning of plant matter, centuries of experimentation led to development of many ways to convert photosynthetic products into light, heat, fuels and chemicals. Some of these processes had been widely commercialized by the 1800s. For example, in the 1850s in the United States, production of alcohol for solvents and lamp oil amounted to 25 million gallons per year. In Europe, internal combustion engines were first designed to run on alcohol, and its production for fuels and chemicals continued into the 20th century. The first commercial system to produce gas from manure was found in Exeter, England in 1895. Used for street lighting, this anaerobic digestion plant was followed by many throughout Europe during times of energy shortages, and millions more small biogas digesters have been built in many developing countries. In the early 1800s, the technique for producing charcoal was turned to converting coal into "town gas" for street lights. The same technique was used in a million small downdraft gasifiers to power vehicles in Europe in the 1930s and 1940s.

Biomass was considered a promising resource because it could be easily converted to a wide range of products including gaseous and liquid fuels, electricity, chemicals and process heat. Today a mixture of old, well-developed technologies co-exists with completely new, advanced techniques for converting biomass to useful energy.

II. TECHNOLOGY DESCRIPTION

Processes for conversion of biomass to heat or to fuels may be divided into five basic categories:

— Direct combustion

— Pyrolysis

— Liquefaction

— Gasification

— Biochemical conversion (anaerobic digestion & fermentation)

Many different technological approaches exist for carrying out the basic conversion process involved. The following will briefly describe the process and some of the technology and Table 27 provides a summary of the basic processes and the high-value energy products they produce.

TABLE 27
BIOMASS CONVERSION TECHNOLOGIES AND PRODUCTS

Process	Products	Premium Energy Resource
Anaerobic Digestion	Methane Carbon Dioxide	Methane
Fermentation/Distillation	Ethanol	Ethanol
Chemical Reduction & Fractional Distillation		High MJ gas or Liquid Hydrocarbons
Pyrolysis	Char, Oils Tars	Methane, Fuel Gas Methanol, Alcohols
Hydrogenation	Oils	Methane, Liquid Hydrocarbons
Direct Combustion	Heat	Electricity

Direct Combustion

Direct combustion is both the oldest and most prevalent contemporary use of biomass. Technology covers a wide range of uses from generation of electricity to industrial production of mechanical energy and furnishing of heat for space heating. Wood is the most common fuel used; however, crop residues, manure, and municipal solid waste (mostly cellulose in its combustible portion) are also commonly used.

Industrial wood combustors may be classified into five types: pile, thin-bed, cyclone, suspension and fluidized bed. After the 1940s, various configurations of thin-bed burners supplanted the Dutch oven pile burners. Efficiency was increased through use of moving, inclined grates. Over time, the spreader-stoker design prevailed, which mechanically throws or pneumatically blows fuel onto the grate. In larger systems, cyclone and suspension systems increase efficiency by keeping the fuel suspended. Fluidized bed systems advanced this concept, using a hot fluidized bed of sand, limestone or other material to incite combustion and provide quick reduction times.

Space heating systems for buildings progressed from free-standing stoves to air-tight designs that provide careful control of combustion and may include insulation, provision of outside air for combustion and heat storage. Catalytic converters may also be added to ensure complete combustion and reduce air pollution. Wood furnaces may be wood-only units, multifuel systems or tandem systems added to existing conventional furnaces.

Pyrolysis, Liquefaction, Gasification

Heating biomass in an oxygen-free or oxygen-deficient environment while minimizing direct combustion produces gas, oil, char and tar. Depending on the products desired, gasification, pyrolysis or liquefaction technology may be used.

Pyrolysis, or the thermal decomposition of an organic material in the absence of oxygen, traditionally has been used to produce charcoal and also may be referred to as retorting, destructive distillation, or carbonization. The slow heating produces roughly equal proportions of gas, liquid and charcoal, but actual output can be varied depending on the feed composition, heating rate, temperature and residence time in the reactor. Higher temperatures and longer residence times promote gas production; while lower temperatures and shorter times produce more liquids and char. A modern process named "fast pyrolysis" minimizes char production to 10% to 20% and can be tailored to produce high yields of either gas or liquid.

Liquefaction technology may operate in direct or indirect systems. Indirect systems involve an intermediate gasification step (see below) followed by catalytic conversion of the product to liquid fuels. Two technological approaches exist for indirect liquefaction conversion. In the first, fast pyrolysis produces an olefin-rich synthesis gas which then can be converted to either a diesel fuel or a gasoline, depending on catalyst and conditions. In the second, gasification leads to production of carbon monoxide and hydrogen and a minimum of hydrocarbons. The resultant synthesis gas can be converted to methanol by proven technology or to mixed-fuel alcohols by new catalysts that are under development. This second indirect process also has been used for peat and coal conversion.

Direct liquefaction technology uses lower temperatures in the reactor to produce partially deoxygenated, complex oils which then are upgraded. Two leading processes are: low pressure pyrolysis to primary vapor followed by catalytic deoxygenation and high-pressure, hydro liquefaction followed by catalytic hydro-deoxygenation. Analysts believe either approach can produce liquid yields from 50% to 70% of the input biomass.

Liquefaction technology commonly is used to produce liquids and chemicals from wood. Many are specialty chemicals, including rosins and turpentines, sulfate and sulfite mill products. In a liquid-fuels-for-energy system, the feedstock can be any high celluloic material that is dried and screened to a fine particle. The feedstock is converted to high temperature gas which then is quenched to produce a liquid. Some gas produced by the system is used to power it, and the char also may be extracted for use.

Gasification refers to the partial combustion of solid carbonaceous fuels to produce a gaseous fuel composed principally of carbon monoxide and hydrogen. Biomass feedstocks perform better than coal because they volatize to 70% to 80% gases compared to 20% to 40% for coal. Furthermore, the biomass char is much more reactive with steam and carbon dioxide in its final conversion to an end product.

Gasification with air produces a low energy gas due to the dilution of the product by the nitrogen present in air. If a higher energy-content gas is desired, such as a carbon monoxide and hydrogen rich synthesis gas, pure oxygen can be substituted for air. Another approach transfers heat into the gasifier through its walls, with heat exchangers or hot sand (also called steam gasification or pyrolysis gasification) so that nitrogen is excluded from the final product. To create a pipeline-quality synthetic natural gas, a two or three stage process is required: gasification—shift conversion— synthesis. The shift conversion uses a catalyst to increase the hydrogen content of the gasification product so that methane can be formed. Appropriate processing and catalysts can be used to create ammonia, methanol or hydrogen.

The major types of gasifiers include: a) updraft (tolerant to fuel form, tar-laden, simple for close-coupled combustion applications); b) downdraft (requires a well-prepared fuel, tar-free, suitable for running internal combustion engines); c) fluid-bed (fuel-form tolerant, moderately clean, adaptable to large sizes); d) entrained flow (requires powdered materials, high throughput).

Biochemical Conversion

Biochemical conversion of biomass involves two technical approaches: anaerobic digestion and hydrolysis fermentation. These technologies depend on biological processes which are complex with many variables affecting output.

In a *fermentation system,* a water slurry or solution of substrate (usually starch and sugars) interacts with micro-organisms and is enzymatically converted to other products. Two of the most common processes yield methane and ethanol. Production of the latter from grain is the most commonly found energy-related technology for fermentation. Fermentable sugars can be obtained directly from certain plants (e.g. sugar cane, beets) or by the hydrolysis of starch or lignocellulose. In the lignocellulose-based process, hydrolysis of cellulose and hemicellulose creates simple flexoses and pentoses which can then be converted by yeasts and bacteria to ethanol or other fuels and chemicals. Acid or enzymes are used to catalyze the hydrolysis reaction of cellulose to sugar, then yeast or bacteria ferment the hexoses to ethanol. The crystalline nature of the cellulose fibres requires pretreatment if the hydrolysis is to be carried out easily. Once the sugars are available, fermentation approaches can produce numerous fuel and chemical products, for example, ethanol, butanol, acetic acid, lactic acid, citric acid, butanediol and acetone. Potential also exists to make aromatic fuels from the lignin fraction of the biomass material.

In an *anaerobic digestion* system, a digester is fed with the biomass feedstock, often with a water slurry. Systems can range from small units found in rural settings that provide gas for light, heat and cooking to large systems used to stabilize sewage sludge for disposal purposes. The first step in the process begins when acidogenic bacteria present in the digestor convert complex organic compounds to lower molecular-weight, soluble products, primarily carboxylic acids. Other microbes coupled with methanogenic bacteria then convert these intermediate products to a gas consisting primarily of methane and carbon dioxide. Residual digested solids, now containing a higher percentage of nitrogen, phosphorus and potassium can be used as fertilizer and animal feed.

Conventional technology operates under non-sterile conditions in large, mixed, anaerobic fermentation vessels at near ambient pressures and at temperatures of 35°C (mesophilic range) or 55°C (thermophilic range) and at reactor residence times from 10 to 20 days duration. Operation in the thermophilic range accelerates the digestion process, reduces investment cost and increases output from these systems. However, thermophilistic digestion seems to be more sensitive to temperature variations and require extra energy in a cold climate to heat the feedstock and to keep the digester at working temperature.

Anaerobic digestors may be batch-type or continuous flow-type systems. The batch-type are the simplest: the biomass is placed in an air-tight, oxygen-free container and left to ferment. Once digestion is complete, the unit is recharged. Three types of continuous flow systems exist. The vertical tank system uses an upright container, and new material is added and mixed as processed waste is withdrawn from the top. Horizontal or plug-flow systems are fed at one end while processed waste is removed from the other. Multiple-phase digestors place the biomass into two containers in sequence for the two phases, optimizing pH and temperature conditions in each. This approach can increase gas yield from 20% to 30%. A wide number of different bacterial cultures is at work in this type of anaerobic digestion and much research work is being done in this field.

III.   APPLICATIONS

Because they convert to fuels that may be substituted for petroleum and natural gas, biomass resources are suitable for a wide range of applications and for use as feedstock for the chemical industry. Biomass can also be used directly to generate electricity and steam.

Agricultural by-products provide two basic types of biomass:
—   dry residues
—   moist residues and wastes.

Dry residues are most valuable when they are rich in cellulose and lignin and their moisture content is below 30% to 35%. Those used most often are grain straw, maize cobs, harvesting residues of such foods as peas and beans, and the prunings from orchards. These biomass resources are most commonly converted by direct combustion to produce thermal energy. Moist residues and wastes come from animal, crop, human, and food-processing wastes.[1] A common conversion technology is anaerobic digestion to produce biogas. This digestion process also produces fertilizers. Energy from agricultural by-products is used often for heating domestic and farm buildings and production of electricity to meet energy requirements of agricultural installations. The residues and wastes may also be collected and used to supplement conventional fuels in the production of industrial process heat and steam.

Wood and forest wastes are commonly converted to energy primarily by combustion. One of the largest applications is for heat, steam and electricity production in the pulp and paper industry. The lumber and wood products industry makes similar use of these resources. Non-forest-product industries that may be located near wood wastes may also rely on them for energy, and utility companies have built wood-fired power plants. Another widely-used application is for residential space heating systems. Wood resources also may be converted to gaseous and liquid fuels with gasification and pyrolysis technologies. Energy output may be used for thermal purposes or to generate electricity.

---

1. Municipal waste is frequently included in discussions of biomass energy conversion, however, it is not included in this review.

Biomass resources grown as energy crops are those rich in sugar or starch, lignin and cellulose or oilseeds. For the most part, the conversion technologies employed to convert such crops to energy are fermentation, hydrolysis/fermentation, and chemical synthesis. Such crops also may be used in combustion systems. The forms of easily transportable energy produced are ethanol, methanol and vegetable oils. The main uses are transportation fuels and raw materials for the chemical industry.

Two countries, Brazil and the United States, have made the most significant commitment to apply biomass to transportation needs. In Brazil, conversion of sugar cane and cassava has been stimulated. In the United States, ethanol is produced primarily from grains.

In summary, biomass resources can meet energy needs for several essential applications: heat for industrial processes and residences, fuels for the transportation and electric power sectors, and feedstocks for energy-intensive chemicals. Along the way, important by-products such as fertilizers and animal feeds are produced. The biomass resources can be used at the point of production to increase the energy efficiency of the producing system, or transported in suitable form to be used in other applications, e.g. utility power, motor fuel and fuel supplements, and chemical feedstocks.

IV.   TECHNICAL STATUS AND ACCOMPLISHMENTS

Status in 1974

Biomass conversion to energy has been used in IEA member countries prior to 1974, but the energy contribution of biomass in the forestry, agricultural and energy sectors was poorly documented. More consideration was given to environmentally sound disposal of logging residues, straw, stover and manure than to energy conversion. Few assessments of the potential energy contribution from biomass existed, in part because of the difficulty of tabulating recovery of wastes in the food and fibre processing industries.

Biomass harvesting technology existed in the early 1970s for both forestry and agricultural biomass; however, it was generally developed to recover only the then economically valuable portion of plants. Collection technology for crop residues was little developed, and only countries short of wood fibre had paid attention to recovery of forestry residues. Large-scale use of wood residue and bagasse to produce process steam was common as a means to dispose of waste. In only a few countries (Scandinavia) were electricity co-generation plants found. Wood stove technology was relatively primative, and few products existed outside the small units for residential applications (10 to 100 kW_{th}).

Processes, such as pyrolysis and gasification, that were well known prior to the 1930s had been revived during World War II but then little developed afterwards. Very few biochemical facilities for production of ethanol (fermentation) or methane (anaerobic digestion) existed in IEA member countries. Environmental concerns had initiated and supported some research in anaerobic digestion beginning in the 1960s. For the biomass conversion to energy field, the early 1970s knowledge base was weak concerning biomass productivity, economically convertible resources of many plant species, development of energy crops, and the environmental impacts of widescale use of bioenergy resources.

Technical Advances

Resource base assessments have been carried out in a majority of countries. Recognition that the total biomass resource is not potentially capturable has given policy makers a better understanding of the economic and environmental limitations that exist for bioenergy systems. Assessments in some countries have shown that 60% of crop residues may not be recoverable for economic and environmental reasons. Policy studies also have indicated that projections of biomass energy potential involve a significant degree of uncertainty about long-term availability of the biomass resources for energy-related uses.

The International Energy Agency has considered wood energy a high priority research area because of its potential for many IEA member countries. The IEA initiated a Forestry Energy (renamed Bioenergy) Agreement to develop voluntary cooperation and coordination in the planning and execution of national biomass research programmes. The two major objectives of the agreement were to encourage information exchange on national activities and to foster co-operative RD&D projects. Forestry energy has been defined by the IEA as the use of short-rotation forest biomass and residues to produce clean fuels, petrochemical substitutes and other energy-intensive products. Although small but significant R&D in wood-based energy technologies is being conducted elsewhere, most bioenergy research and development in this field is being pursued in IEA member countries.

Continuing improvements in forestry growth and production technologies and wood harvesting and conversion are enhancing the value of wood as a source of renewable energy. Some of the key research has concentrated on expanding the potential of several fast-growing species (populus, salix), analysis of nutrients required, and development of harvesting techniques and the technology needed to increase economic recovery and yields. The energy plantation concept has also stimulated research in sugar crops, such as Jerusalem artichokes, and in oil-bearing plants, such as milkweed and kochia. Fresh and salt water growth of algae and other types of aquatic plants for energy and chemical feedstocks has also been actively researched. Associated research on environmental aspects of large-scale energy "farming" has begun to identify maintenance and protective measures required for these options to be economically sound on a long-term basis.

In the area of recovery of forestry biomass, major advances have been made in harvesting systems, both for whole trees and for logging residues. Original designs developed for fibre recovery evolved into machines that harvest brush and stands. Commercial versions of this type of equipment are now available and entering the market. Harvesting equipment for agricultural residues and wastes has also been further developed, particularly for straw and stover recovery. Harvesting and transportation technology for aquatic biomass resources are less advanced.

Despite the fact that combustion technology could have been considered mature in the 1970s, research and engineering have increased the efficiency and reduced environmental impact of these systems. Residential wood stove technology has increased in efficiency by up to 30% and development of catalytic convertors and designs to enhance secondary combustion have lowered the release of particulate matter and by-products. Despite these advances, the effect on air quality from wide-spread use of residential systems still remains a concern. In some jurisdictions, authorities have required use of catalytic converters in wood stoves. In a few countries with peat moss, large-scale production of energy through harvesting has advanced from insignificant energy contributions to a level of economic importance, particularly in Finland.

Gasification technology of biomass has been proven at the demonstration level for boiler retrofits, crop drying, and in internal combustion engines operating electric generators. New developments in the use of semi-dry forest biomass and residues have led to further refinements in gasification technology and new demonstration projects. The technical feasibility of producing medium energy gas and methanol has been demonstrated. This technology also has progressed to a commercial state, with some companies developing systems for the market.

Technology for production of syngas has been advanced to pilot plant stages in a number of countries. A primary goal is the production of methanol for transportation fuel. Liquefaction technologies also have been advanced, but they have not yet shown the economic promise of other biomass conversion options. Basic research into pyrolysis for gasification of biomass to produce oils and high-value organic liquids has revealed important factors involved in the conversion process and created an improved base for further technological advances.

Anaerobic digestion technology has been significantly advanced. Feedstock research has shown that all land, aquatic and marine biomass species are suitable for digestion and can be pretreated to promote it. Giant brown kelp has been identified as a promising feedstock. Technology has been improved from simple low-cost batch and plug flow digestors to advanced concepts such as the two-phase process developed in the United States. Such advances have reduced retention times, increased yields and net energy production. Commercial development of these technologies is advancing in the food processing industry and in large-scale agribusiness.

Ethanol production in the United States and Brazil has reached large-scale proportions, and the technical feasibility of ethanol to supplement or supplant gasoline for transportation has been well demonstrated. Major research accomplishments include improved efficiency of the conversion process and expansion of the feedstock base. In many IEA countries, active research on conversion of ligno-cellulosic materials (e.g. wood, straw) is expanding the potential of these materials as a feedstock for liquid fuels. Liquid fuels research on oil seeds has demonstrated that plant oils are suitable to run diesel engines or extend conventional diesel fuels.

In summary, research and development advances have occurred across a broad front. Existing technology with commercial potential has been improved and acceptance of biomass conversion to energy expanded. Already commercial systems have been improved, and existing applications to control and dispose of wastes have been successfully augmented to produce energy. Wide-ranging basic research on conversion chemistry has laid important groundwork for continued advances. Biological sciences have contributed significantly to understanding of the resource base and the potential of many species of plants to become energy-producing products.

The IEA Biomass Conversion Technical Information Service has been supporting this research via its bibliographical service which produces abstracts of publications. This service is available to IEA members and other subscribers doing research and development in the biomass field.

Commercial Status

Some types of biomass energy systems are commercial in many countries. In some cases commercial status already existed in the 1970s. For others, it has developed to significant levels in the last decade. There are a number of biomass conversion options that have not yet made any commercial progress. There are also technologies that were used many years ago which have been improved but which still remain uneconomic under current competitive conditions. The commercial market penetration that has occurred varies by country, but in most cases it extends beyond agriculture and the forestry industries to the residential, industrial, and utility sectors.

In Finland, Ireland and Sweden, peat, forest and agricultural resources provide 13% to 19% of the primary energy supply. Other IEA member nations such as Belgium, Denmark and New Zealand are at the other extreme with biomass resources meeting only a fraction of a percent of energy demand. In Denmark, however, straw is substituting for 15% of coal consumption in co-fired applications, and the expected potential is near 30%. Austria, Canada, Norway and the United States meet approximately 3% to 4% of their current energy needs from biomass. The scale of the contribution in the United States, some 63.6 million Mtoe in 1983, exceeds the total energy requirements of the Netherlands.

Table 28 presents data on the total energy supplied by biomass in 1983 for those countries participating in the IEA forestry energy programme.

TABLE 28
ENERGY CONTRIBUTION FROM BIOMASS TO IEA COUNTRIES IN 1983

Country	% of Energy from Biomass	Energy Demand (Mtoe)	Energy from Biomass (Mtoe)
Austria	4	25.8	1.03
Belgium	0.2	39.5	0.08
Canada	3	211.3	6.34
Denmark	1	16.6	0.17
Finland	17	24.4	4.3
Ireland	13	8.5	1.11
New Zealand	0.4	12.0	0.05
Norway	4	25.2	1.01
Sweden	13	48.6	6.32
Switzerland	1.1(1.6)	25.6(16.3)	0.27
U.S.	3(4)	1727.5	63.6
U.K.	0.3(1985)	183.7(1985)	0.60

Forest residues contribute a very large share of bioenergy to IEA member countries. In the United States, residential wood stove sales increased from 235 000 per year in 1973 to a peak of 2.1 million in 1981. Industrial use of wood energy is prevalent in the forest products industry, which in IEA countries meets 30% to 55% of its energy needs through biomass conversion. In the United States, the largest market developed to date, about 40% of biomass consumption is in the residential sector, and 60% in the industrial and utility sectors. Utility-owned power plants total 131 MW_e, and another 850 MW_e of power plants are independently owned and selling power to utilities.

Utility adoption of biomass technology for electric power plants has begun to expand. Conversion of municipal solid waste leads such development; however, agricultural waste, forest residue, and animal manure systems are also being utilized or considered for commercial projects.

Denmark, Finland and Sweden have constructed a number of biomass-fueled district heating plants. The market in Sweden has expanded from two plants totaling 9 MW_{th} of capacity prior to 1980 to 38 plants with a total capacity of 550 MW_{th}. The Swedish government provided a 25% subsidy in 1983 and 1984 to further market development of peat and wood biomass systems. The current subsidy is 1%.

In 1982, peat accounted for 2% and wood-based biomass for 17.3% of energy supply in Finland. Hundreds of commercial peat-fired plants exist, the largest with a capacity of 265 MW_{th}. Finland also has developed commercial peat gasification plants in MW scale, and several are in operation or under construction for district heating. Ireland has also developed its peat resources and more than 50 000 ha of bogland are currently in production. Seven peat-fired power stations are generating 18% of Ireland's electricity. Two plants are rated at 45 MW$_e$; several use wood as a secondary fuel source.

Although important research efforts have been undertaken to develop short rotation crops, no significant commercial penetration has occurred. Direct combustion biomass systems are not as simple as liquid or gas fossil fuel burners, and capital cost per energy unit produced remains higher compared to gas and oil-fired equipment. Competitiveness is restricted therefore to relatively low-cost feedstocks. This leads to consideration of densified biomass fuels which can be used in modified conventional boilers.

Little commercial penetration is evident for many of the gasification processes under development. An early market application has been demonstrated in Finland and Sweden where oil is conserved by burning gasified biomass in lime-kilns at wood processing plants. In sum, gasification is not yet really commercialized. Furthermore, few companies, if any, are attempting to commercialize pyrolysis or liquefaction systems.

Commercial utilization of agricultural wastes is expanding in on-farm systems to produce heat, steam, and in some cases electricity. Typical applications in the United States are fueled by beef and dairy cattle manure and waste from food processing plants (potatoes and soy beans). In the global picture, the contributions are small.

Conversion of biomass to ethanol has been commercialized in Brazil and the United States. Brazil increased its production from 640 million litres in 1975 to 8 billion litres/year by the mid-1980s. The goal for the 1987-1988 harvest year is 14.3 billion litres. Brazilian production has amounted to an equivalent of 170 million barrels of gasoline.[1] In the United States, anhydrous ethanol production rose from 75 million litres in 1979 to 1.6 billion litres in 1984. Maize was the dominant feedstock, accounting for 54% of production. Other feedstocks included: other grains 22%; molasses 13%; and food waste 11%.[2] In Europe, little commercial production of ethanol as a liquid fuel has developed. Both the Brazilian and United States industries rely on government subsidies for their economic competitiveness.

Because costs of biomass feedstocks vary significantly from region to region, any presentation of energy costs from biomass energy conversion technologies must only be imprecise. The technologies themselves are inhighly varied stages of commercial readiness, some fully mature, others still in conceptual stages. Even when demonstrations of newer technologies exist, it

---

1. Finneran, K., Innovation: Development and Status of the Renewable Energy Industries 1985. Solar Energy Industries Association, Washington, D.C., 1986, p. 111.
2. Ibid., p. 200.

remains difficult to furnish precise cost figures because few data are available on closely monitored systems in commercial configurations. Finally, the diversity of sites, the varied economic relationships, and the value of trade-offs involved to a certain economy make any transnational comparisons of cost data somewhat hazardous.

V.   ENVIRONMENTAL CONSIDERATIONS

Tests show that vehicles running on alcohol or vegetable oils produce less air pollution than systems running on oil-based fuels. In some countries, lead emissions from gasoline are being reduced by using alcohol blends with gasoline. Figure 24 compares the emissions from an ethanol/gasoline blend with gasoline. The biomass-derived fuels, however, produce more aldehydes or formaldehydes than gasoline. The impact of this type of air pollution is not completely understood.

FIGURE 24
INFLUENCE OF FUELS ON EXHAUST EMISSIONS, 24°C

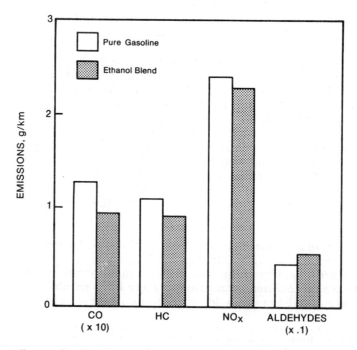

Source: J.R. Allsup and D.B. Eccleston, Department of Energy, USA -Ethanol/Gasoline Blends as Automotive Fuels - 1979

From an environmental perspective, the long-term impact of intensive culture practices needs to be further analyzed. In a few years' time research should be providing clearer answers to issues involving water use patterns, loss of nutrients by removal of residues and the impact of herbicides and pesticides. With regard to direct combustion, technical fixes in stove design promise to mitigate the air quality effects of wood stoves in densely populated regions. For larger systems, proven technology already exists to control particulate and organic emissions. Sulfur emissions are not a problem, and in fact blending of biomass with coal can reduce total emissions to acceptable standards.

Pyrolysis and liquefaction have potentially harmful waste streams of polynuclear aromatics and other organics. Their concentrated nature suggests confinement and recyling back to the reactor for destruction. Biochemical conversion can release compounds that can pollute air and water unless the pollutants are systematically controlled. Distillery waste-waters from alcohol fuels production can be a serious problem; however, several disposal options now exist, including bio-oxidation, anaerobic digestion and fertilizer-irrigation. Costs for clean-up should be low to moderate, and the waste can be handled by existing environmental control technologies.

VI.   TECHNOLOGY OUTLOOK

For some biomass technologies, competitiveness has been achieved, while for others, such competitiveness lies in the future, dependent on technological improvements as well as increases in conventional energy prices. Real costs can only be determined when conditions in existing markets for biomass, whether agriculture or forestry based, are studied in detail and trade-offs and net costs are fully understood. In general, the recent decline of petroleum prices has made many biomass energy options unattractive from the standpoint of cost of delivered energy. Nonetheless, direct combustion is thriving economically in many niche markets, and its market should continue to expand. Anaerobic digestion in agricultural settings is also proving its economic viability, and where wood and residues are available locally at favorable spot prices, gasification systems appear marginally economic.

It appears certain that technological progress will favor the increased competitiveness of biomass energy systems. More difficult is determination of which technological options have the greatest likelihood of a major breakthrough that would significantly reduce costs. The timing for such developments should be fairly rapid because the technology lends itself to low-cost testing of concepts. Furthermore, the whole field is related to biotechnological research where rapid progress is being made.

Advances in genetic research and genetic engineering could have profound impacts on the agricultural production of biomass specifically for energy purposes. More efficient conversion of nitrogen could lessen need for fertilizers, lowering costs to produce biomass feedstocks. Hybridization of various species could also increase yields through adjustments in protein and carbohydrate compounds.

The cost of ethanol produced from lignocellulosic feedstocks (biochemical conversion) is a factor of 1.5 to 3 times more expensive than current costs for ethanol produced from grain and sugar crops. Research being conducted in IEA member countries shows great promise for reducing these costs to below the level of costs for grain-produced ethanol by the end of the century. Advanced cellulose conversion technology with application of modern acid hydrolysis and enzymatic processes is rapidly evolving and will reduce ethanol production costs. Furthermore, processes for utilization of the hemicellulose and lignin fraction of the feedstocks are expected to advance biofuel conversion processes to a competitive position with petroleum-based fuels and chemicals.

The environmental impact of biomass production for energy purposes should require careful consideration. For biomass energy crops, the impacts are most associated with soil quality and resulting changes in ecosystems. These effects may be mitigated by prudent harvesting. Other effects, notably from combustion and those involving toxic compounds, can be handled with known technology and techniques. Overall, the effects will vary for each production and conversion technology and will also vary depending on the feedstock. The environmental issue is important and serious review of ways to minimize impacts on land, water and air quality are required for regions where biomass is or will be used for energy production.

Co-operative research and development and information exchanges are addressing biomass conversion issues related to both thermal and biochemical conversion and the cross-cutting issues common to both genetic processes. The IEA has just formulated a new plan of co-operative research under the production and conversion annex of its biomass programme. Tables 29 and 30 provide details.

In the area of feedstock production, genetic research is concentrating on poplars, willows, alders and conifers. Other research is devoted to evaluation of the stock, propagation and banking of superior genetic stock for safe distribution and use. Various countries are participating in collection and distribution of promising parent species. Goals in production biology are related to better understanding of the relation of nutrient and water supplies and coppicing to growth. Strategies to enhance the economic viability of short rotation forestry plantations are being developed, including research on fertilizer and irrigation management regimes. The morphological and physiological basis of coppicing will be studied, the best methods determined, and instructions produced. Research on operation of plantations will aim to improve stand management and harvesting technology. Indeed, recent advances in wood fuel harvesting equipment are being introduced commercially in Denmark, Finland, Norway and Sweden.

TABLE 29
IEA BIOMASS PRODUCTION ANNEX PROJECT ACTIVITIES 1986

Area	Activity	Leader, Country	Nature of Activity	Results
Genetic Improvement and Biotechnology	Poplar Ex.	Steenackers, BEL	Coop. R & D	Superior Stock
	Alder Eval.	Hall, USA	Coop. R & D	Superior Stock
	Willow Br.	Zsuffa, CAN	Coop. R & D	Superior Stock
	Con. Br. Strat	Zsuffa, CAN	Info. Ex	Breeding Strategies
	Joint Eval.	Perttu, SWE	Coop. R & D	Improved Evaluation Criteria
	Cell Cult.	Cheliak, CAN	Coop. R & D	Improved Techniques
	Biotech. Dev.	Ericsson, SWE	Coop. R & D	New Techniques
Production Biology	Nutr. Cyc.	Miller, UK	Coop. R & D	Fertilizer Prescription
	Coppice	Ferm, FIN	Coop. R & D	Improved Coppice
Operations	Prod. Tech.	Mattson, USA	Tech. Dev.	Improved Technology
	Economics	Lothner, USA	Info. Transfer	

TABLE 30
IEA BIOMASS CONVERSION ANNEX PROJECT ACTIVITY 1986

Activity	Lead Country	Lead Institution	Project Leader
Thermal Conversion			
Direct liquefaction	SWE	Energetics AB	Bjorn Kjellstrom
Combustion safety	NOR	SINTEF	Erling Oesterboe
Thermal conv. conf.	FIN	To be determined	
World-wide Data Base	UK	University of Aston	A.V. Bridwater
Biochemical Conversion			
Pretreatment of lignocellulosics	NZ	Forest Research Institute	Keith Mackie
Conv. of C-5 sugars	NZ	Massey University	Graham Manderson
Alcohol fuels	CAN	Forintek Canada Corp	John Saddler
Anaerobic digestion	CAN	Canviro Consultants, Inc	A. R. Stickney
General Conversion			
Municipal solid waste conversion	UK	AERE Harwell	Chris Dent
Environmental Issues			
Combustor emissions	NOR	Center for Industrial Res.	Christel Benestad
Voluntary Standards	CAN	Solar Energy Research Institute	Thomas Milne
Operating Agent	CAN (Temp)	Battelle Pacific Northwest Lab.	Don J. Stevens

Economic evaluation of short rotation plantation concepts needs to be further refined and recommended methodologies disseminated to cooperating countries.

For thermochemical conversion processes, co-operative research activities should improve the efficiency and effectiveness of thermally producing liquid, gaseous and solid fuels from biomass. Combustion processes for heat and electricity production will also be included. With the exception of direct liquefaction and novel pyrolysis systems, thermochemical conversion systems are either commercial or achieving maturity through demonstrations. The European Economic Communities are investing in syn-gas production, while Canada, Sweden and the United States have their own programmes emphasizing syn-gas and methanol production from biomass and peat.

While wood gasification has little application under current conditions and with current capabilities, the prospects of increased usage are likely to improve as gasifiers improve. Gasification research needs to address problems that interrupt smooth operation of such systems. These include feeding systems, handling ash, and control and removal of tars. Fast pyrolysis and direct liquefaction technologies are least understood. Improvements are expected in heat transfer, yields, catalytic upgrading and introduction of feedstocks to the reactor. Experiments on very high heat rates and use of reactive atmospheres have indicated 40% to 60% increases in yields in bench scale tests. Important R&D breakthroughs are required to improve product yield and quality.

For biochemical conversion processes, emphasis is being given to conversion of lignocellulosic feedstocks via fermentation and digestion. Pretreatment of feedstocks to increase reactivity in conversions is also being investigated. Technological advances are anticipated in the following areas:

— Micro-organisms for use in anaerobic digestors
— Selection techniques for microbial strains
— Genetic engineering of superior microbes, yeasts, fungis
— Pretreatment processes for lignocellulosic feedstocks
— Catalytic processing of lignins to liquid fuels
— Fermentation to convert pentoses to ethanol.

Improvements are expected to contribute to the competitive production of significant quantities of liquid and gaseous fuels and energy intensive petrochemical substitutes by the year 2000.

Anaerobic digestion technology for producing methane has advanced, and research in basic biochemical and microbiological reactions, material durability and process engineering should increase system efficiency. Technical advances in pretreatment of feedstocks (e.g. municipal solid waste and agricultural residues) should also increase yields. Advances have demonstrated that by-products such as fertilizers will improve with gains in system efficiency.

Ethanol production through fermentation has gained significantly in efficiency, particularly with regard to energy balances. Grain systems in the United States show a 1.5:1 advantage, including energy required for growing feedstocks. Brazilian sugar conversion ratios are near 9:1. New technology, based on bacterial screening and other micro-organic substrates, promises important advances in direct fermentation hydrolysis. Success of steam explosion techniques has established a new technological base for breaking down cellulose structure.

Further advances are expected in the area of aquatic crop production and conversion. The technological goals include development of varieties that provide sustained high yields and large amounts of lipids and design of innovative production and harvesting systems. Genetic engineering offers the potential of creating species tailored to tolerate saline water regimes. The high yields produced by algae in small-scale experiments have not been duplicated in larger scale tests and more research is required to expand the energy potential of such resources.

Outlook

Better inventory and potential supply projections will be forthcoming in the next few years. Studies have shown that tabulating the physical extent and biological productivity of biomass resources is not a reliable indicator of economic potential.

Norway, the single European net energy exporter, is conducting biomass R&D to increase utilization of its biomass resources by the year 2000. Finland, already relying on biomass for nearly 20% of its energy supply, has a national policy of increasing energy self-sufficiency from 32% to 35% through utilization of forest and peat resources. In Sweden, it is fully expected that biomass will increase its already important contribution to energy supply by 2000.

Canada has vast forestry resources which will be increasingly used for energy supply. Bioenergy was supplying about 6% of Canadian energy requirements in 1983. The commitment to development through support for R&D and direct incentives for bioenergy projects is expected to lead to an energy contribution from biomass sources near 8% by 2000. Austria, meeting almost 4% of its needs in 1983 with bioenergy, expects the contribution to increase threefold by 2000 through expanded use of fuelwood, bark, sawmill residues, straw, biogas and energy farming. The government in Denmark has estimated that wood, agricultural wastes and straw have the potential to supply from 5% to 9% of total energy supply by the end of the century.

In New Zealand the potential for major expansion of wood supply at conventional rotation ages is very limited. The government expects that short rotation forestry plantations could, however, significantly increase the biomass potential by 2000. Some preliminary estimates project energy contributions up to 40% but many factors (e.g. export potential for paper and pulp versus energy) must be taken into account. The forestry sectors in other IEA countries such as Belgium and Ireland are more limited in potential. Belgium for example imports 50% of its timber. Ireland, with only 5% of its land under forest, has the least amount of wood resources among EEC countries. In these and other countries that must import wood to satisfy domestic demand, ultimate wood energy potential will depend to a large degree on opportunity costs.

In 1983, biomass resources were reported to account for approximately 3% of total energy supply in the United States. The largest consumers are the pulp and paper and lumber industries, which are self-generating more than 50% of their needs. Very large potential is forecast for expanded wood energy production in the United States. The current national energy plan estimates that if total energy demand increased by roughly 30% by 2000, the contribution of biomass resources could range from 12% to 17%. These estimates do not include micro- or macro-algae or advanced herbaceous crops cultivated as energy feedstocks. Their contribution is not expected to become significant until after 2000.

In sum, the outlook for biomass energy contributions are favorable in every country. Where commercialization has occurred already, it is expected to expand. Where biomass resources are under-utilized, it is expected that their contribution will increase. A survey of IEA member countries in 1983 indicated that ten nations expect at least 5% of their primary energy supply to come from biomass by the year 2000 and that for the IEA countries in total, biomass contributions could increase from a total of nearly 1.6% in 1980 to 7.4% in 2000. It must be realized, however, there is a difference between projections based on national intentions and estimated economic potential, and that comparisons of biomass contributions among IEA member countries can only be general in nature.

VII. MAJOR FINDINGS

Biomass energy sources are making significant contributions to energy supply in many IEA member countries. Technological status ranges from mature to experimental, and market development is quite uneven. Substantial market penetration has occurred in residential and industrial sectors where biomass resources are readily available. As with any other renewable energy resource, the economic potential from biomass is quite site specific. The capability to predict potential resource contributions has improved dramatically in the last decade.

Many countries already realize significant energy contributions from biomass and look to further exploit this resource. Most countries have completed the phase of R&D on biomass that involves resource assessment. The ultimate economic potential of biomass conversion to energy will inextricably involve national policies regarding forests, agriculture and land use. Furthermore, environmental concerns pose serious issues and will also shape the extent and pace of bioenergy development.

Scientific progress across a broad spectrum of biomass technologies has resulted in significant accomplishments over the last decade and has improved the economic potential of many technical options. Advances include:

— Review of hundreds of herbaceous plants for hydrocarbon content and screening of still more experimental hybrids;

— Comprehensive and coordinated research on short rotation woody crops that confirm higher growth rates for controlled plantations than in natural settings and supports emphasis on realizing the economic potential of this approach;

— Anaerobic digestion technology has been enhanced by several key innovations and should continue to expand in its early markets;

— Research on conversion of lignocellulosic material to ethanol shows promise, but more R&D advances will be required to advance this approach;

— Biomass conversion technologies have been successfully applied to conversion of municipal solid waste in many countries, and this field is going to expand rapidly in several IEA member countries;

Thermochemical conversion research in gasification, liquefaction and pyrolysis indicate promising new processes; however, liquefaction and pyrolysis require more R&D advances to approach competitiveness. Significant gains are still realizable for the technologies already commercialized or near commercialization; consequently immediate expansion of biomass potential does not have to rely on more advanced concepts and technologies.

A key barrier to greater biomass utilization remains its relative lack of economic competitiveness. Capital investment and operating costs are generally higher than for conventional systems. Government incentive programmes have successfully stimulated introduction and furthered commercialization of various biomass energy conversion options.

VIII. BIBLIOGRAPHY

1. Gilliusson, Rolf, *National Bioenergy Programmes in Sweden*, Swedish University of Agricultural Sciences, Garpenberg, 1984.

2. Gilliusson, Rolf, *National Research and Development Programmes in Member Countries of the IEA Forestry Agreement Area "Harvesting, On-Site Processing and Transport"*, Swedish University of Agricultural Sciences, Garpenberg, 1984.

3. Gilliusson, Rolf, *Review of National Programmes 1984 Sweden*, Swedish University of Agricultural Sciences, Garpenberg, 1985.

4. Jaycor, *Review of U.S. Biomass Energy Activities for IEA*, Alexandria, Va, USA, 1984.

5. Jones, Noel, *Bord Na Mona*, No. 59 of the Irish Environmental Library Series, Dublin.

6. Mattsson, Jan E. and Nilsson, Per P. (editors), *Proceedings of the International Conference, "Harvesting and Utilization of Wood for Energy Purposes"*, Garpenberg, 1981.

7. McMullen, J.T. and Strub, A.S., (Prepared for the Commission of the European Communities), *Achievements of the European Community Second Energy R&D Programme*, Brussels, 1984.

8. National Energy Administration of Sweden (Acting as Operating Agent for IEA project Forestry Energy), *IEA Forestry Energy Annual Report 1983*, Sweden, 1984.

9. Wilhelmsen, Gunnar, *The Norwegian R&D Programme on Energy from Biomass*, The Agriculture Research Council in Norway, 1984.

10. Klass, Donald A., *Handbook of Energy Technology and Economics*, John Wiley and Sons, New York, 1983, p. 712.

11. *Environmental Effects on Electrictiy Generation*, OECD, 1985.

12. Solar Energy Industries Association, *Energy Innovation: Development and Status of the Renewable Energy Industries, 1985*, Washington, D.C., 1986.

13. Renewable Energy Institute, *Annual Renewable Energy Technology Review: Progress through 1984*, Washington, D.C., 1986.

14. Organisation for Economic Co-Operation and Development, *Biomass for Energy: Economic and Policy Issues*, Paris, 1984.

CHAPTER FIVE: GEOTHERMAL ENERGY

A. NATURE OF THE GEOTHERMAL ENERGY RESOURCE

Geothermal energy is the thermal energy stored in rocks and fluids within the earth. It is estimated that approximately 10% of the world's land mass contains accessible hydrogeothermal resources that could theoretically provide hundreds of thousands of megawatts of energy for many decades. Strictly speaking, geothermal resources are not renewable on a human time scale. Reservoirs can become depleted in a matter of decades and thus require careful phasing, conservation, and reservoir management policies. On the other hand, geothermal does not share the intermittent quality of many renewable energy sources and can deliver energy as required.

Depending upon the thermodynamic properties of the extracted fluid, resources are classified as low-enthalpy systems (where temperatures range from 30°C to 150°C) or high-enthalpy systems (above 150°C). A slightly different categorization defines low-temperature systems as those below 90°C, moderate-temperature systems as ranging between 90-150°C, and high temperature systems as those above 150°C. Low- and moderate-temperature systems are used for direct heat. The higher temperature sources are used primarily for electricity production. In order to tap geothermal resources commercially, the heat content of geological formations must be concentrated in a restricted volume at accessible depths, and temperatures must fall within a useable range.

In stable geological zones, the temperature of the earth increases with depth by a geothermal gradient of 3°C for every 100 meters. This means that at a depth of 2 km the temperature of the earth is about 70°C, increasing to 100°C at a depth of 3 km and so on. However, in some places geophysical activity allows hot or molten rock to approach the earth's surface, creating pockets of higher temperature resources at more accessible depths.

Hydrogeothermal energy is distributed widely but unevenly across the Earth. High enthalpy geothermal fields occur within well-defined belts of geologic activity, often manifested as earthquakes, recent volcanism, hot springs, geysers and fumaroles. The geothermal belts are associated with the margins of the earth's nine tectonic or crustal plates and are located mainly in regions of recent volcanic activity or where a thinning of the earth's crust has taken place.

One of these belts rings the entire Pacific Ocean and penetrates across Asia into the Mediterranean area. Hot crustal material also occurs at mid-ocean ridges (e.g. Iceland and the Azores) and interior continental rifts (e.g. the East African rift). The shaded areas in Figure 25 indicate geothermal regions associated with plate margins. Areas of low grade resources are not fully depicted.

Figure 25
WORLD WIDE LOCATION OF MAJOR GEOTHERMAL SYSTEMS

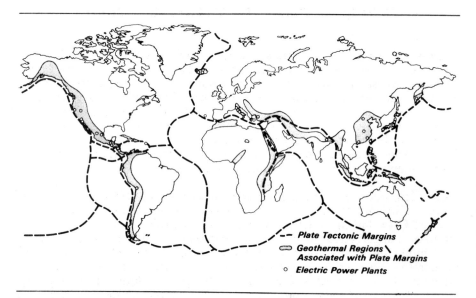

Source: United States Geothermal Technology, Equipment and Services for Worldwide Application, DOE/ID-10130

High enthalpy, recoverable resources available for power generation far exceed the development to date. Many countries are believed to have potential in excess of 100 000 MW_e which would fulfill a considerable portion of their electricity requirements for many years.

Low enthalpy resources are more abundant and distributed more widely. They are located in many areas such as deep sedimentary basins along the Gulf Coast of the United States, in Western Siberia, and in certain areas of Central and Southern Europe.

Classification of Resources

There are four types of geothermal resources: hydrothermal, geopressured, hot dry rock, and magma.

Hydrothermal resources, which are the most commonly used at the present time, contain hot water and/or steam trapped in fractured or porous rock at shallow to moderate depths (from approximately 100 m to 4500 m). Hydrothermal resources are categorized as vapor-dominated (steam) or liquid-dominated (hot water) according to the predominant fluid phase. Temperatures of hydrothermal reserves range from 90°C to over 350°C, but roughly two-thirds are estimated to be in the moderate temperature range (150°C - 200°C). The highest quality reserves contain steam with little or no entrained fluids, but only two sizable, high quality dry steam reserves have been located to date— Larderello in Italy and the Geysers field in the United States.

Geopressured geothermal resources are hot water aquifers containing dissolved methane trapped under high pressure in sedimentary formations at a depth of approximately 3 km to 6 km. Temperatures range from 90-200°C although the reservoirs explored to date seldom exceed 150°C. Geopressured reserves are rare, and the only major resource area identified to date is in the Northern Gulf of Mexico Region where large reserves are believed to cover an area of 160 000 km².

Hot Dry Rock resources are accessible geologic formations that are abnormally hot (above 150°C) but contain little or no water. Hot Dry Rock resources are found all over the world at different depths, but extracting their energy is difficult. They are generally more accessible in young volcanic centres.

Magma is molten rock at temperatures ranging from 700-1200°C. Magma chambers represent a huge potential energy source, the largest of all geothermal resources, but they rarely occur near the surface of the earth. Extracting magma energy is expected to be the most difficult of all resource types. The accessible depths are thought to be between 3 000 and 10 000 m. This resource has not been developed as yet, and many questions remain to be answered.

Hydrogeothermal Resources in the IEA Region

On the whole, since western Europe lies within a tectonically stable area of old and relatively rigid continental crust, the majority of IEA countries have primarily low enthalpy resources. A younger crust with associated recent volcanism stretches over the Mediterranean region, however, giving Italy, Greece and Turkey areas of high temperature geothermal resources. Portugal has high-temperature resources and has operated a power plant on the island of San Miguel in the Azores.

Having about 10% of the active volcanoes in the world, Japan is famous for its volcanic activity. Considering only the shallow geothermal energy resources up to 2 000 meters in depth, Japan is estimated to have potential resources of approximately 30 GW of electricity-generating capacity for thirty years which is equivalent to about 51 Mtoe per year. New Zealand also has major geothermal resources.

The United States is endowed with all of the various types of geothermal energy. It possesses the largest known steam field; liquid-dominated hydrothermal systems with temperatures in excess of 150°C have been identified at fifty-three sites; and low-temperature fluids are widely available. In addition, a very large accessible geopressured resource base is located along the coast of the Gulf of Mexico; Hot Dry Rock sources have been identified at a number of locations; and accessible magma bodies are believed to exist at quite a few sites.

B. GEOTHERMAL ENERGY TECHNOLOGIES

I. INTRODUCTION

Although geothermal energy has had a long history of use in applications such as therapeutic hot baths, space and water heating, and agricultural growth stimulation, it was only in 1904 that the power of natural geothermal steam was first harnessed to produce electricity by Prince Piero Ginori Conti at Larderello, Italy. During the very early decades of its use for this purpose, growth was slow because of cheap competing sources of electric power. Only those sources which were easy to exploit such as the big geysers in California, United States and some liquid-dominated geothermal resources in Japan and New Zealand were developed to the commercial power stage.

II. TECHNOLOGY DESCRIPTION

Geothermal energy is used in two ways - for electricity generation or direct use of heat. The temperature of the extracted fluids greatly influence the use to which the resource is put. Generally, electricity generation is the most attractive use of geothermal resources because it can be transported cheaply. This is important since reservoirs are often located at some distance from energy markets. The economics are usually more attractive for resource temperatures over 200°C, but new technologies are being developed to utilize more moderate (150-200°C) temperature fluids.

Low-to-moderate temperature resources, which are much more plentiful than high-temperature resources, are most commonly used for direct-heat applications. This is also an excellent application of geothermal energy since direct applications tend to be technically simple, thermodynamically efficient, highly adaptable, and cost-effective. However, location of the reservoir (proximity to the use) becomes an important issue since heat cannot be transported economically over even moderate distances. From a practical perspective, direct heat applications should probably be sited within 1 km of the wellhead.

Of the geothermal extraction techniques, hydrothermal energy is the most advanced and cost-competitive and the only one presently being used commercially. Magma, geopressured and Hot Dry Rock systems are still in experimental stages; although the latter two types have been successfully penetrated by wells, and energy extraction has been experimentally verified. Hot Dry Rock exploration has been carried out in the United Kingdom, Japan, France, Germany, the Union of Soviet Socialist Republics and the United States. The latter is the only country currently investigating geopressured and magma resources.

Exploration and Extraction

For the most part, geothermal resources are located and extracted using technologies adapted from the oil and gas industries. Their application to geothermal technology has only been partially successful, however. Efforts are therefore underway to develop improved methods and materials which can deal with the temperatures, salinity and abrasive nature of the geothermal environment and to develop better procedures for forecasting geothermal reservoir performance.

Geothermal hydrothermal development begins with exploration to locate and confirm the existence of a reservoir with economically exploitable temperature, depth and volume. Most known reservoirs were discovered from surface manifestations such as hot springs, but exploration now relies increasingly on techniques such as geologic maps, gravity meters for assessing the density of the rock, electrical methods, seismics, geochemical thermometers, sub-surface mapping, and temperature and heat flow experiments.

This step is followed by exploratory drilling and production testing to establish reservoir properties. If a suitable reservoir is confirmed, field development follows. The geometry and physical properties of the reservoir are modeled; changes in reservoir fluids and rock are analyzed; long-term behavior of the reservoir is predicted by numerical simulation; and the siting of production wells and injection wells for disposal of spent brines is determined.

Other types of geothermal energy have special requirements in the exploration phase. For example, the forces that drive fluids from geopressured brine reservoirs are much different from those in conventional oil and gas reservoirs and require a special technology for forecasting geopressured reservoir performance. Better sensing techniques besides seismic methods are needed for exploring magma deposits.

Hydrothermal drilling technology requirements and costs increase as the geothermal environment becomes hotter, deeper and more difficult to drill. Recovery of geopressured energy requires high-pressure technology and the use of heavy drilling muds. Hot Dry Rock requires the drilling of deep wells with high directional accuracy in very hard rock and creating artificial heat exchange fractures through which fluid can be circulated, entering and leaving through a pair of deep wellbores. The basic concept in Hot Dry Rock technology is to form a man-made geothermal reservoir by drilling deep wells (4 000-5 000 m) into high-temperature, low-permeability rock and then forming a large heat exchange system by hydraulic fracturing. Injection and production wells are joined to form a circulating loop through the man-made reservoir, and water is then circulated through the fracture system. Figure 26 illustrates this geothermal technology.

Successful magma drilling technology has not been established. Magma technology will require special drilling technology when the high temperature (over 400°C) deposits are encountered to deal with the interactions of the drill bit with molten rock, the effects of dissolved gases, and mechanisms of heat transport in molten magmas.

After the geothermal fluids are extracted, their thermal energy can be used directly or converted to electricity.

III. APPLICATIONS

Direct Heat Applications

The abundant low and moderate temperature hydrothermal fluids (less than 180°C) can be used as direct heat for space and water heating, for industrial processes, and for agricultural applications. Some of the major uses have been for heat and hot water for hospitals, schools and churches; district heating systems for groups of buildings (the predominant use); greenhouse heating; warming fish ponds in aquaculture; crop drying; and for various washing and drying applications in the food, chemical, and textile industries. See Figure 27 for an indication of the range of direct-use applications and their temperatures.

In direct-use geothermal systems, fluids are generally pumped through heat exchangers to heat air or a liquid; although the resource can be used directly if the solids content is low. These systems are the simplest applications, using conventional off-the-shelf components.

Conversion (Electricity Generation)

If temperatures are high enough, the preferred use of geothermal resources has been for generation of electricity which could either be fed into the utility grid or be used to power industrial processes on site. Since the geothermal resource is relatively constant, i.e. it does not encounter daily or seasonal fluctuations, it can be used for baseload power production.

FIGURE 26
HOT DRY ROCK (HDR) CONCEPT

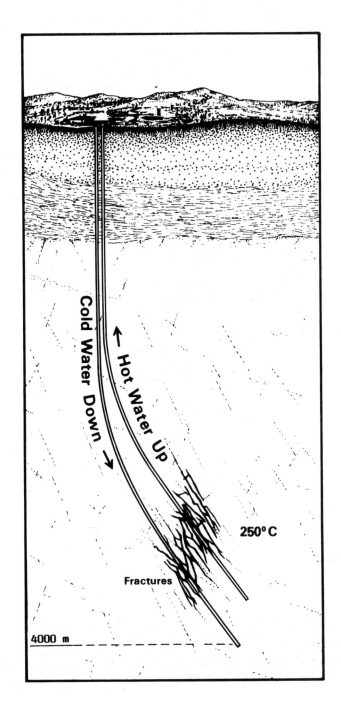

FIGURE 27
DIRECT USE APPLICATION TEMPERATURES

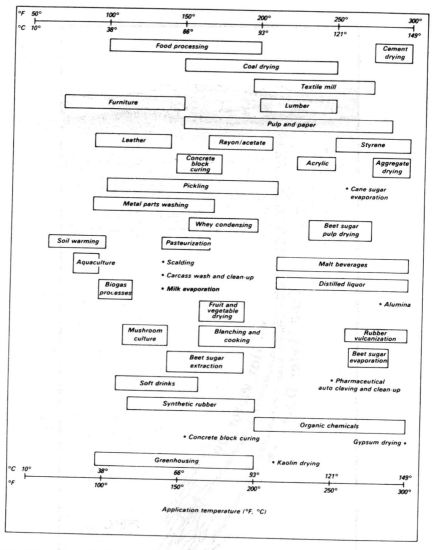

Source: *United States Geothermal Technology:* Equipment and Services for Worldwide Applications. United States Department of Energy (DOE/ID-10130), 1985, p. 31.

There are several types of energy conversion processes for generating electricity from hydrothermal resources: dry steam and flash steam systems, which are traditional processes, and binary cycle and total flow systems, which are newer processes with significant advantages. Figure 28 shows both a traditonal steam power plant and a binary cycle power plant.

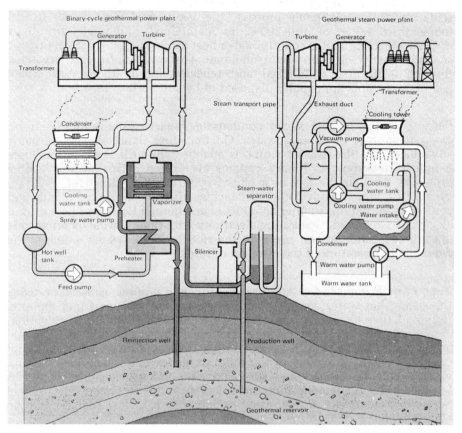

Source: New Energy Development Organization, Japan 1984.

Dry Steam Plants are used to produce energy from vapor-dominated reservoirs. Steam is extracted from the wells, cleaned to remove entrained solids and piped directly to a steam turbine. This is a well-developed, commercially available technology, with typical plants in the 35-120 MW_e capacity range.

Flash steam plants are used to produce energy from liquid-dominated reservoirs which are hot enough (typically above 200ºC) to flash a large proportion of the liquid to steam. Single-flash systems evaporate hot geothermal fluids into steam and direct it through a turbine. In dual-flash systems, steam is separated from hot water and fed into a dual-inlet turbine, while the remaining hot water is flashed into steam and then directed through the turbine. In both cases the waste brine is either used for cooling water or is reinjected into the reservoir. This technology is economic at many locations and is being developed in sizes of 10-55 MW_e.

Binary Cycle plants are appropriate for liquid-dominated resources which are not hot enough for efficient flash steam production. They are also useful for saline brines that cannot be flashed because of the resulting deposition of scale. In this conversion process, the geothermal hot water, in the temperature range of 150-200°C, is maintained under pressure by a down-well pump and is sent through a heat exchanger where it vaporizes a secondary working fluid (e.g. isobutane). The working fluid is expanded through a turbine, condensed and reheated for another cycle. Spent geothermal fluids are usually disposed of by sub-surface injection.

Because of the lack of steam condensate, binary systems must rely on external sources of cooling water. On the other hand, they offer many advantages, not the least of which is enabling the utilization of moderate temperature resources. Binary systems also avoid corrosion and scaling problems through the use of a closed-cycle system. Moreover, they are capable of higher conversion efficiencies than flash steam plants and consequently require smaller amounts of geothermal fluids per heat unit of electricity generated. Generating electricity with binary cycle technology is the primary candidate for use with the Hot Dry Rock resource, due to the moderate temperature of the circulated fluids.

Total Flow Prime Movers, another experimental process, is based on using the steam, hot water, and pressure of geothermal resources (i.e. the total resource), thereby eliminating energy losses associated with the conventional method of flashing and steam separation. These systems usually channel a mixture of steam and hot water into a rotating conversion system and capture the kinetic energy of the mixture to power an electric generator. Work on total flow technology was conducted under the auspices of an IEA Implementing Agreement and resulted in a prototype system (the Helical Screw Expander) which was tested in Italy, Mexico, New Zealand and the United States.

Hybrid electric plants supplement geothermal heat with biomass or fossil fuel. Such plants may be appropriate for reservoirs which are too cool to generate electricity economically from the hot water alone but which can play a fuel-saving role.

Utilization of geopressured resources, which contain recoverable methane and whose reservoirs are under very high pressure, requires additional steps in the energy conversion process for extracting the methane. After methane extraction, conventional technology may be used, including binary plants, hybrid electric plants, or impulse turbines to capture mechanical energy or heat exchangers for direct heat applications.

Potential *magma energy applications* include possibility of electric power production and of using the chemical reactivity and/or heat of the magma to evolve hydrogen from water or synthesize gas from water/biomass mixtures.

V.   STATUS AND ACCOMPLISHMENTS

Status in 1974

Hydrothermal. The only countries which were exploiting geothermal energy for electricity production on an industrial scale in 1974 were Italy, Japan, New Zealand and the United States which had a total installed capacity of about 770 MW$_e$. Electricity from geothermal energy has been generated in Italy for more than eighty years, primarily from dry steam plants. High temperature fluids from steam-dominated and water-dominated fields were used directly from wells or from flash cycles in condensing turbines. The residual water was discharged into rivers or evaporating ponds and the uncondensable gas discharged into the atmosphere through ejectors or multi-stage compressors.

Direct Heat. There were relatively few examples of direct use of geothermal heat in industry or agriculture other than geothermal greenhouses in a few countries. Italy and New Zealand utilized geothermal resources in the pulp and paper and chemical industries, respectively. Geothermal district heating was also in use in Hungary, Iceland and the United States.

Geopressured. The United States' geopressured geothermal research began in 1974 with the objective of determining whether the geopressured brine aquifers of the United States Gulf Coast could be exploited as a major domestic source of energy. Initial research was limited to geological studies of the Gulf Coast sedimentary basins and preliminary investigations into methods for using the energy in the brines for power generation and direct uses.

Hot Dry Rock. The United States' Hot Dry Rock programme was launched in 1971 with funding of research at the Los Alamos National Laboratory where the concept of extracting heat from Hot Dry Rock was developed in 1970. By 1974, two exploratory holes had been drilled at the Fenton Hill, New Mexico experimental site that confirmed the basic principles of the technology.

Magma. While the potential of magma as an alternative energy source had been recognized for some years, it was not until mid-1973 that Sandia National Laboratories proposed active investigation. A conference on the utilization of volcanic energy was held in 1974 in Hawaii which attracted international attention to the concept of magma use and increased the level of scientific interest.

Accomplishments

Significant advances have been made since 1974 in developing effective geothermal technology and in determining the engineering and econoomic feasibility of exploiting geothermal energy systems. Emphasis has been placed on power production using the liquid hydrothermal resource since power production with dry steam has been commercially viable for several years. Hot Dry Rock technology is still in the experimental stage and has been the subject of intensive research in a number of countries including a collaborative project under IEA auspices conducted in the United States. Considerable research effort has been devoted to geopressured and magma resources, but only by the United States.

Hydrothermal Research The major elements of hydrothermal research have been reservoir definition, brine injection technology, heat cycle research, emissions abatement, permeability enhancement, reservoir investigations, geothermal materials development and deep scientific drilling.

Considerable progress has been made in refining geological, geophysical and geochemical technologies as a result of the research of the past 10 years. Exploration technologies designed especially for hydrothermal reservoir evaluation have been developed. Reservoir definition and engineering have been advanced by new methods for characterizing fractured reservoirs and realistically predicting their response to production and injection. Reservoir predictive models can be used today with limited success. A number of reservoir confirmation projects have been conducted, and the state of the art of reservoir characterization has been advanced through the development of instruments for well testing. Improved drilling systems and components are now available, but costs are still too high and research is continuing.

Extensive materials development has been undertaken in response to serious problems encountered with material failures. Many materials typically used for mechanical components suffer thermal degradation on exposure to the hot fluids, and the chemicals present in the fluid create costly problems with corrosion and scaling.

The materials research has resulted in a number of solutions to these problems through high temperature elastomers and polymer concrete, leak-tight metallic seals, high temperature down-hole cable, steels for improved drill bits and pitting-resistant alloys. In addition, standardized fluid sampling and analysis procedures have been developed, and precipitators/clarifiers and scale inhibitors to handle salinity and corrosive brines are now commercially available.

Power conversion systems utilizing very hot fluid resources have reached maturity, and several flash steam generating plants are operating or under construction. Heat conversion research has included operation of pilot scale binary cycle units for moderate temperature fluids, and several plants are now operating successfully and competitively. Research now centers on larger binary units and on those for use with highly saline brines.

Thus, major strides have been taken in improving the viability of hydrothermal technology. Numerous technical advances have been made, but further improvements are needed, particularly in the areas of exploration techniques and resource assessment, reservoir behavior, improved drilling techniques, and binary cycle conversion.

Geopressured Resource research has been directed toward determining the economic feasibility of extracting the resource as well as gaining a better understanding of the quantity, productivity and longevity of the reservoirs. For the past ten years, the United States Department of Energy has funded extensive long-term field tests on a number of wells on the Gulf Coast of Louisiana and Texas as well as laboratory research.

This research has established that the quantity of geopressured brines contained in the Gulf Coast is very large. The large quantity of methane saturated in the brines can be extracted from the brine by a simple and economical gravity separation technique. It is believed that United States petroleum industry drilling and production technology can be successfully adapted to geopressured brine production. Researchers have also concluded that, because of their relatively benign nature, spent brines may be disposed of by injection into saline reservoirs at relatively shallow depths.

A major problem revealed during field tests has been calcium carbonate scaling which can severely restrict brine flow in a well. A scale inhibitor utilized in a recent experiment has apparently been successful in achieving an effective solution, inasmuch as the test well has been producing about 30 000 barrels of brine per day for more than six weeks with no sign of scaling.

The temperatures of the brines have been the cause of some concern. While they range between 120°C and 205°C in situ, the temperatures in test wells seldom exceed 150°C which is marginal compared to other geothermal resources. However, advances in binary conversion, as well as the possibility of co-production of methane, may make the low temperature resource profitable. A hybrid geopressured conversion test unit is presently being constructed in the United States.

Hot Dry Rock. A major long-term Hot Dry Rock (HDR) research effort has been underway since 1972 at Fenton Hill in New Mexico. The objective of the project, which is carried out by Los Alamos National Laboratory, in collaboration with German and Japanese scientists in the IEA Hot Dry Rock Implementing Agreement is to determine whether an industrial-scale artificial reservoir can be created and operated economically for commercial power generation. Early experiments proved that energy can be extracted at rates of up to 5 MW_{th}. The objective of the current experiments is to determine whether a reservoir of sufficient size and longevity can be made which would attract private investment in commercial power generation from Hot Dry Rock.

A new reservoir at Fenton Hill was created in 1983 by flow pumping of over 20 million litres of water into the rock under high pressure. The rock temperature in the reservoir is about 240°C. The target design for the system is a lifetime of at least 10 years with less than 20% temperature drawdown. In June 1985, the two wells were successfully connected. Long term heat extraction experiments are now underway; early results showed an energy extraction rate of 9 MW_{th} at an exit temperature of 190°C. The final evaluation of this promising experiment is expected to take place in 1988.

A complementary approach to the Fenton Hill work has been adopted in the HDR experiments at the Camborne School of Mines in Cornwall, in the United Kingdom. The aim of this work, funded by the United Kingdom Department of Energy with some support from the Commission of the European Communities, has been to understand the complex processes which take place when rocks at depth under stress are fractured. Differences between the Fenton Hill and United Kingdom projects include British use of explosive fracturing and high viscosity fracturing fluids. The Cornwall reservoir is less than 100°C.

A successful interconnection was achieved between the first two wells in Cornwall, but the hydraulic connection was poor, causing flow impedance and water loss. As a result, only about 60% of the water was recovered. In an attempt to improve the reservoir connection, a third well was deviated into the fracture at a different angle. Recent tests have shown a great improvement in circulation characteristics. Heat extraction tests have been conducted, and researchers are considering working at greater depths.

The Soviet Union has also begun a HDR experimental programme with the drilling of a 4 300 m well in the western Ukraine. France, Germany and Japan are other countries with projects developing this technology.

Magma. In the ten years since interest first developed the possibility of extracting heat from shallow magma sources, the scientific feasibility of doing so has been established. During a United States Department of Energy-funded multi-year research project conducted by Sandia National Laboratories, experimental boreholes were drilled to 74 m into buried molten rock at temperatures up to 1 100°C, and heat extraction experiments were successfully demonstrated at the Kilauea Iki crater in the Hawaii Volcanoes National Park. The approach used was to cool the lava to solidify it before drilling rather than directly contacting the liquid rock. The technology required to drill into molten or semi-molten magma at depth is being evaluated since the western continental United States has a number of relatively young volcanoes that probably overlie magma chambers. A site in Long Valley, California has been selected for intensive investigations of drilling and completion technology, characterization of the magma environment, and energy extraction studies.

Commercial Status

Substantive market penetration has thus far occurred only in hydrothermal technology. During the last decade, the greatest growth has occurred in the electricity sector, but considerable activity has also taken place in direct use applications.

Market Penetration - Hydrothermal Direct Heat. At the end of 1984, the total installed capacity for direct use of geothermal energy worldwide was 7 100 MW$_{th}$, which provided about 24 000 GWh of energy. The primary users of direct geothermal heat and 1984 capacity are found in Table 31. The total installed capacity in 1975 was about 3 100 MW$_{th}$; thus the growth in direct uses in the past decade has been approximately 8.5% per year.

TABLE 31
CAPACITY OF DIRECT USE GEOTHERMAL PLANTS IN OPERATION - 1984
(For countries having a capacity above 100 MW$_{th}$)

Country	Flow Rate Kg/s	Power MW	Energy GWh	Load %
China	3540	393	1945	56
France	2340	300	788	30
Hungary	9533	1001	2615	30
Iceland	4579	889	5517	71
Italy	1745	288	1365	54
Japan	26101	2686	6805	29
New Zealand	559	215	1484	79
Romania	1380	251	987	45
Soviet Union	2735	402	1056	30
Turkey	1355	166	423	29
United States	1971	339	390	13
Other	1965	142	582	47
Total	57803	7072	23957	39[1]

1. Based on total thermal power and energy

Source: J. Gudmundsson, *Direct Use of Geothermal Energy in 1984*. 1985 International Symposium on Geothermal Energy.

In Italy, a saving of approximately 0.30 Mtoe per year is expected by 1990 as a result of direct hydrothermal energy increasing to as much as 1 Mtoe by the year 2000. Eight new district heating systems and commercial greenhouses are operational there. Extensive studies in Turkey, which has a number of greenhouses and space heating projects already in operation, indicate a potential of at least 31 000 MW$_{th}$.

Japan is the world's largest user of direct heat, with the use of geothermal waters for balneology being the major application there. Other uses are in aquaculture, greenhouse heating, district heating, agricultural applications and industrial processes.

There are over 200 direct-use projects on line in the United States providing an estimated 400 GWh annually. The major end uses are space and water conditioning projects and district heating systems, which account for 50% of direct heat production. Other important applications include commercial fish farms, commercial greenhouses, and industrial process heat (IPH) projects, including copper processing, vegetable drying and industrial process uses. Although potentially very economically attractive, geothermal heat for processes is limited by the difficulty in relocating a commercial enterprise to a geothermal site. The growth of direct use applications in the United States may also be affected by the expiration of the federal residential energy tax credits.

New Zealand was a pioneer in the use of direct geothermal energy. Applications there include district heating, space cooling and an industrial process heat plant.

Market Penetration - Hydrothermal Electric

Worldwide installed geothermal electric capacity amounted to approximately 4 800 MW_e in 1985. Table 32 shows the countries with installed power plants as of that year, the number of units, type of plant and capacity.

TABLE 32
GEOTHERMAL POWER PLANTS ON-LINE AS OF 1985

Country	No. Units	Type(s)[+]	MW_e
United States	56	DS,1F,2F,B	2022.11
Philippines	21	1F	894.0
Mexico	16	1F,2F	645.0 *
Italy	43	DS,1F	519.2 *
Japan	9	DS,1F,2F	215.1
New Zealand	10	2F	167.2
El Salvador	3	1F,2F	95.0
Kenya	3	1F	45.0
Iceland	5	1F,2F	39.0
Nicaragua	1	1F	35.0
Indonesia	3	DS,1F	32.25
Turkey	2	1F	20.6
China	12	1F,1F,B	14.32*
Soviet Union	1	F	11.0
France (Guadeloupe)	1	2F	4.2
Portugal (Azores)	1	1F	3.0
Greece (Milos)	1	1F	2.0 *
Totals	188		4763.98

+ DS = dry steam; 1F,2F = 1- and 2-flash steam, B = binary
* Includes plants under construction and scheduled for completion in 1985

Source: Ronald DiPippo, *Geothermal Electric Power-State of the World, 1985*, International Symposium on Geothermal Energy.

The growth of geothermal electric power capacity has accelerated rapidly since the late seventies, evidenced by a 16.5% annual growth rate. This figure is expected to reach 5 500 MW_e by the end of 1986.

Italy. Italy was the first country to produce electricity from geothermal steam, at Larderello in 1904. At the end of 1985, the installed capacity for electricity production in Italy amounted to about 520 MW_e, nearly all from dry steam plants (See Table 33). The more challenging task of using the liquid-dominated resources is being addressed with a 4.5 MW_e pilot plant at Latera. Unfortunately, Italy's geothermal energy utilization has been hampered by the depletion of its steam fields. In 1984, about 3 400 metric tons per hour of steam were used, essentially the same figure as in 1975. Nonetheless, energy production has increased as a result of plant modernization to increase efficiency. Moreover, ambitious plans have been launched to explore the entire length of the Apennines, and new high temperature fields have been discovered near Larderello. The National Energy Plan forecasts that installed capacity will reach 600-800 MW_e by 1990.

TABLE 33
GEOTHERMAL POWER PLANTS IN ITALY

Plant	Year	No. Units	MW(1)	Type (2)
Larderello 2	n.a.	4	58.0	C
Larderello 3	1969	5	113.0	C
Gabbro	1969	1	15.0	C
Casteinuovo	n.a.	4	50.0	C
Serrazzano	n.a.	5	47.0	C
Sasso Pisano	n.a.	1	3.5	NC
Sasso Pisano	n.a.	2	15.7	C
Lago	n.a.	3	33.5	C
Monterotondo	n.a.	1	12.5	C
Radicondoli	1979	2	30.0	C
San Martino 1	1980	1	9.0	C
Lagoni Rossi 1	1960	1	3.5	NC
Lagoni Rossi 3	1981	1	8.0	C
Molinetto 2	1982	1	8.0	C
Travale 2	1973	1	15.0	NC
Travale 2	1946	1	3.0	C
Piancastagnaio	1969	1	15.0	NC
Bagnore 1	1945	1	3.5	NC
Bagnore 2	1945	1	3.5	NC
La Leccia	1983	1	8.0	C
Latera	1984	1	4.5	NC
San Martino 2	1985	1	15.0	C
Pianacce	1985	1	15.0	C
Bellavista	1985	1	15.0	C
Unspecified	1985	1	15.0	C
Total			519.2 operational	

(1) Plant totals.
(2) All plants are dry steam types except Latera which is 1 - Flash;
   C = condensing; NC = noncondensing turbine.

Source: Geothermal Electric Power, The State of the World - 1985, by Ronald DiPippo, Southeastern Massachussets University, U.S.A., 1985.

Japan. Since the inauguration of the Matsukawa geothermal power plant in 1966, nine power stations have come into operation with a total generating capacity of 215 MW$_e$ (See Table 34). With the exception of Matsukawa, which utilizes steam-dominated resources, all the plants use hot water geothermal fluids. In Japan, two binary-cycle pilot plants of 1 MW capacity each were completed in 1979. Another 55 MW$_e$ unit is to be constructed, and several plant expansions are under consideration.

TABLE 34
GEOTHERMAL POWER PLANTS IN JAPAN

Name of Company	Name of Power Station	Maximum Capacity (MW$_e$)	Year of Commission	Running Factor (%) (4)	Load Factor (%) (5)
J.M.C. (1)	Matsukawa	22	October 1966	94	87.2
Kyushu E.P. Co.	Otake	12.5	October 1967	94.8	84.2
Mitsubishi Metal Co.	Onuma	10	June 1974	95	89
E.P.D.C. (2)	Onikobe	12.5	March 1975	95.3	70.1
Kyushu E.P.Co.	Hatschobaru	55	June 1977	95.1	89.3
Tohoku E.P. Co., J.M.C.	Kakkonda	50	May 1978	95.1	90.3
Suginoi Hotel	Suginoi (3)	3	August 1981	97.8	80.1
Hokkaido E.P. Co. Donanjinetsu Energy Co.	Mori	50	Nov. 1982	94	57
Daiwabou Kanko Co.	Kirishima Kokusai	1	Febr. 1984	95	63
Total	9	215.1	—	—	—

(1) Japan Metals and Chemicals Co., Ltd
(2) Electric Power Development Corporation
(3) Designated for use in a hotel compound
(4) Running Factor and Load Factor are based on 1982 figures
(5) Load Factor = average electricity production per calender year/maximum capacity x 100

Source: Local Energy, Vol. 3, No. 2, New Energy Foundation, Tokyo, 1983

According to a 1983 long-term forecast for energy supply and development, the contribution of geothermal energy to the overall energy supply in Japan is expected to increase from 0.34 Mtoe in 1982 to about 1.3 Mtoe in 1990. By the year 2000, the contribution is expected to increase ten-fold.

New Zealand. Between 1959 and 1963, thirteen units totalling 167 MW$_e$ were installed at Wairakei, the oldest geothermal plant operating on a liquid-dominated reservoir. That plant has experienced some decline in reservoir pressure and the loss of high-pressure steam. Wairakei has maintained a remarkable record of performance over its lifetime: an average load factor of 88% over the fifteen years from 1968 to 1982. It should be noted that Wairakei was originally designed for use with a heavy water production plant that was never built and has the most elaborate flashing arrangement of any geothermal plant. Thus, the excellent record of reliability and high performance is all the more impressive. A new double flash power plant is now under construction at Ohaaki with a capacity of 116 MW$_e$. Other sites have also been discovered that may someday be exploited. Table 35 shows the development of geothermal power plants in New Zealand as of 1985.

TABLE 35
GEOTHERMAL POWER PLANTS IN NEW ZEALAND

Plant	Year	MW$_e$		Status
Wairakei:				
Unit 1	1959		11.2	Operational
Unit 2	1958		6.5	Dismantled
Unit 3	1959		6.5	Dismantled
Unit 4	1959		11.2	Operational
Unit 5-6	1962	2x	11.2	To be installed at Ohaaki
Unit 7-8	1959	2x	11.2	Operational
Unit 9-10	1960	2x	11.2	Operational
Unit 11	1962		30.0	Operational
Unit 12-13	1963	2x	30.0	Operational
Kawerau	1969		10.0	Operational
Ohaaki:				
Unit 1	1988	2x	11.2	Under construction
		2x	46.9	
	TOTALS:		167.2	Operational
			283.4	Operational, or under construction

Source: Geothermal Electric Power, The State of the World 1985, by Ronald DiPippo, Southeastern Massachussets University, U.S.A. 1985

Turkey. In addition to the 20 MW$_e$ geothermal power plant commissioned in January 1984 in the Denizli-Kizildere geothermal field, continued efforts are underway to assess the geothermal energy potential in other promising regions. High temperature reservoirs have been discovered in two regions and construction of a second power plant (with two 55 MW units) is being considered by the Turkish Electricity Authority.

United States. The United States is the world's largest producer of electricity from geothermal energy. As of 1985, the United States had about 2 000 MW$_e$, installed generating capacity in five western states. A geothermal industry is emerging that reaches beyond the impressive dry steam resource at The Geysers and is making major gains in developing the hot water resource.

United States public utilities have plans for increased capacity for utility-owned plants and increased purchases of power generated by non-utility owned plants.

A number of successful pilot plants have been constructed which have provided useful data for subsequent facilities. An industry/Department of Energy pilot project at Niland, California, established the feasibility of using flash steam systems for electric power production from high temperature, high salinity resources. Magma Power Company has constructed an experimental 10 MW$_e$ dual binary plant at East Mesa, California for

Table 36
GEOTHERMAL POWER PLANTS AT THE GEYSERS, UNITED STATES

Plant	Year	MW$_e$		Status
PG&E Geysers:				
Unit 1	1960		11	Operational
Unit 2	1963		13	Operational
Unit 3	1967		27	Operational
Unit 4	1968		27	Operational
Unit 5-6	1971	2x	53	Operational
Unit 7-8	1972	2x	53	Operational
Unit 9-10	1973	2x	53	Operational
Unit 11	1975		106	Operational
Unit 12	1979		106	Operational
Unit 13	1980		133	Operational
Unit 14	1980		109	Operational
Unit 15	1979		59	Operational
Unit 16	1985		114	Under construction
Unit 17	1982		114	Operational
Unit 18	1983		114	Operational
Unit 19	n.a.		55	Preliminary planning
Unit 20	1985		114	Under construction
Unit 21	1988		140	Advanced planning
Unit 22	n.a.		114	Preliminary planning
Unit 23	n.a.		114	Preliminary planning
Unit 24	n.a		114	Preliminary planning
Wild Well	1985		1.2	Advanced planning
NCPA 2	1983		110	Operational
SMUDGEO No. 1	1983		72	Operational
Bottlerock	1985		55	Operational
UXY 1	1984		80	Operational
NCPA 3	1985	2x	55	Under construction
Modesto GEO	n.a.		110	Preliminary planning
South Geysers	n.a.		55	Advanced planning
SMUDGEO No. 2	1987		55	Preliminary planning
CCPA No. 1	1988		55	Under CEC review
CCPA No. 2	n.a.		55	Preliminary planning
	Totals		1 792	Operational[2]
			2 660.2	Oper., u.c. or planned

(1) All units are dry steam type except Wild Well which will be a binary plant.
(2) Includes plants under construction and scheduled for completion in 1985.

Source: Geothermal Electric Power, The State of the World - 1985, by Ronald DiPippo, Southeastern Massachussets University, U.S.A., 1985.

utilization of a hot water resource and a 34 MW$_e$ high salinity flash plant at the Salton Sea. Both plants are operating continuously and selling power to Southern California Edison. Union Oil and Southern California Edison have completed two 10 MW$_e$ single-flash demonstration plants.

A 45 MW$_e$ proof-of-concept binary plant, the largest in the world to date, has been recently completed at Heber, California. The purpose of this plant, jointly-funded by Department of Energy and the San Diego Gas and Electric Co. and other utilities, is to validate the viability of binary cycle technology for large plants using resources in the temperature range 150-200°C.

Until recently, geothermal market penetration in the United States centered around The Geysers, the world's largest geothermal field, which has been a successful commercial venture for two decades. The Geysers has a total of 19 plants operating, all of the dry steam type, which have a combined capacity of 1 790 MW$_e$. Pacific Gas and Electric has had the longest and most extensive experience at The Geysers, having operated there since 1960. The company currently has 1 137 MW$_e$ of capacity at that location which, in 1983, produced 6 billion kWh of electricity or 8.4% of the total power available for sale to the utility's customers. Three additional plants are under construction, and four more are planned. Several municipal utilities are also tapping The Geysers' resources. The Northern California Power Agency has one 110 MW$_e$ plant on-line and another under construction, and the Sacramento Municipal Utility District has completed a 72 MW$_e$ plant. See Table 36 for a list of all projects in The Geysers field.

In the last few years, a number of power plants utilizing liquid-dominated resources have begun operating or are under construction. Six plants were in operation in the Imperial Valley, California, at the end of 1985 and included units of the binary, single-flash and double-flash types totalling 160 MW$_e$ capacity. See Table 37 for a list of the projects in the Imperial Valley. The total installed capacity for the rest of California, Nevada, Oregon and Utah is estimated to be about 250 MW$_e$. United States utilities and field developers plan to have about 3 300 MW$_e$ of geothermal power on-line by the end of 1992.

Other IEA Countries

Greece has begun to tap the large potential of Milos and other islands lying along the volcanic arc of the Aegean Sea. New Zealand is the site of the world's first commercial hot water geothermal power plant.

Non-IEA Countries

The Philippines ranks second only to the United States in geothermal electricity production with 894 MW$_e$ installed and another 150 MW$_e$ planned for the next few years. Mexico has moved into third place with a total of 710 MW$_e$ capacity operational or under construction. The govern-

TABLE 37
GEOTHERMAL POWER PLANTS IN UNITED STATES (IMPERIAL VALLEY, CA)

Plant	Year	Type		MW$_e$	Status
East Mesa:					
B.C. McCabe No.1	1979	Binary		12.5	Operational
Magma Unit 2	n.a.	Binary		25.0	Planned
Magma Unit 3	n.a.	Binary		25.0	Planned
ORMESA (Ormat)	1986	Binary	26x	0.77	Under construction
Salton Sea:					
Geothermal Electric					
project (Union/SCE/					
SPLC/MPC)	1982	1-Flash		10.0	Operational
Vulcan Power Plant					
(Magma/SCE)	1985	2-Flash		34.5	Under construction
Niland (NPN					
Partnership)	n.a.	2-Flash		49.0	Planned
Niland Geothermal					
Energy Program					
(Parsons)					
Phase 1	1986	2-Flash		38.6	Under construction
Phase 2	1988	2-Flash		31.4	Planned addition
Heber:					
Binary Demo Plant	1985	Binary		45.0	Under construction
Flash Plant (HGC)	1985	2-Flash		49.0	Under construction
North Brawley	1980	1-Flash		10.0	Operational
Westmorland	1988	Binary		15.0	Planned
South Brawley (CU 1)	n.a.	Flash		49.0	Planned
		Totals		161.0	Operational *
				219.62	Op. or u.c.
				414.02	Op., u.c. or planned

* Includes plants under construction and scheduled for completion in 1985

Source: Geothermal Electric Power, The State of the World - 1985, Ronald DiPippo, Southeastern Massachussets University, U.S.A., 1985

ment hopes to achieve 2440 MW$_e$ by the year 2000. Indonesia is a relative newcomer with only 30 MW$_e$ on line, but there are plans for an additional 745 MW$_e$ by 1994. El Salvador has been generating power from geothermal sources since 1975. A total of 95 MW$_e$ are now on line there.

V.   ENVIRONMENTAL ISSUES

The major environmental issues associated with geothermal energy are airborne emissions, liquid effluents, noise, induced seismicity and subsidence. Since geothermal reservoirs can have a range of characteristics, the incidence, type and severity of environmental impacts from geothermal development are very site-specific as well as process-specific.

Airborne Emissions. Geothermal fluids are quite varied, ranging from potable quality water to fluids with complex mixtures of dissolved gases and solids. As these latter types of fluids are withdrawn from a reservoir and processed to produce electricity in flashed steam facilities, reduction in temperature and pressure may cause volatilization and subsequent release of undesirable gases that do not condense at atmospheric temperatures and pressures. In particular, atmospheric release of hydrogen sulfide (H_2S), which is present in nearly all high-temperature geothermal fluids, has been of special concern. At high concentrations, H_2S is toxic and corrosive; however, the primary problem has been its annoying odor which can be detected by 20% of the population at a concentration of just 0.002 parts per million by volume.

Substantial progress has been made in the development of highly effective emission abatement systems. Both the Stretford scrubbing process and the newer Dow RT-2 System which produces no toxic wastes have been successfully utilized in the United States at The Geysers. All plants at The Geysers are now in compliance with California emission limits, and the new plants are emission-free. The flash plant at the Salton Sea reservoir in the Imperial Valley also incorporates effective abatement measures.

Airborne emissions are not a problem with closed-loop binary cycle systems as long as geothermal fluids are kept at pressures high enough to prevent volatilization of gases.

Liquid Effluents. Disposal of spent hydrothermal and geopressured fluids, which may be highly saline and toxic, is one of the more important issues affecting development of the technology. The salinity of the water means that care needs to be taken to prevent contamination of surface and sub-surface waters.

The two basic methods of liquid disposal are surface disposal and injection. Surface disposal is only possible if the fluid is pure enough to avoid adverse environmental consequences or if the brines have been treated to prevent contamination of surface and ground waters. In other instances, the fluid must be disposed of by subsurface injection. Injection of the spent brines into the production reservoir can be advantageous since this process recharges the system and prevents subsidence.

Land Subsidence. Another impact of geothermal development can be subsidence, or sinking of land, caused by the withdrawal of large quantities of fluids from underground reservoirs. An example of this problem can be seen at the Wairekei, New Zealand power plant where a depression covering 65 km^2 has developed, with a maximum rate of subsidence of about 0.4 m/year. Injection of the withdrawn fluid is the primary mitigation method, but its general effectiveness is still uncertain. A plant in the Imperial Valley, California is being monitored by the United States Department of Energy in an effort to study the impact of brine injection.

Induced Seismicity. Many hydrothermal reservoirs are in regions that experience frequent natural seismic activity. It is believed that the withdrawal and injection of geothermal fluids might increase the frequency of microseismic events. At Latera, Italy, reinjection was clearly the cause of seismic activity, but this activity was localized and of low intensity. Field tests indicate that the use of lower injection pressures avoid the danger of induced seismicity.

Noise. Geothermal noise is caused by drilling at the wells, by venting of high pressure steam during development of a new borehole, and by ground vibrations at uncontrolled wells. Rock mufflers, although expensive, can greatly reduce noise caused by venting of steam to the atmosphere. Other noise sources can be controlled by conventional techniques.

VI. **TECHNOLOGY OUTLOOK**

Key Work Currently Underway

Hydrothermal R&D: The following have been identified as the key remaining technical and economic impediments to greater development of hydrothermal reservoirs:

— Lack of confidence in reservoir potential assessment techniques

— Uncertain behavior of injected fluids

— Inadequate drilling technology

— Needed reductions in the cost of binary cycle plants

The research underway to address these problems is described below.

Reservoir Potential Assessment. The characteristics and behavior of geothermal reservoirs are not yet predictable with the certainty necessary to induce large investments in the use of the resource. This difficulty is compounded by the fact that the nature and quality of the resource varies from reservoir to reservoir. Consequently, various amounts of geoscientific study and costly confirmation drilling are required for each reservoir. To alleviate this problem, improved models are being sought for prediction of reservoir production capacity. Better surface and subsurface exploration and mapping techniques are also being developed.

Brine Handling. Efforts are continuing to develop economically and environmentally acceptable technologies for handling of brines in the energy conversion phase and for sub-surface injection of spent brines. The objective is to develop models to analyze fluid migration, to predict thermal and chemical effects, and to optimize injection well placement and reservoir operations. Studies are also being conducted in brine treatment to better understand these chemically complex fluids and develop better methods for coping with them.

Drilling Technology. Work in this area is designed to achieve improvements in hard rock penetration technologies, which includes all aspects of drilling, logging, well completion, and other extraction operations. Rock penetration mechanics, advanced borehole instrumentation for temperatures over 250°C, and new drilling and coring concepts are being emphasized.

Binary Conversion. Attention is being focussed on ways to increase the performance of large binary geothermal power plants and reduce cooling water requirements in order to make moderate temperature resources viable for electric power generation. Monitoring of the 45 MW_e Heber binary plant continues to provide information on equipment and reservoir performance and the economic and environmental acceptability of binary cycle plants.

Geopressured R&D. Research is underway to determine the characteristics and projected longevity of typical geopressured reservoirs under production conditions. An experimental power generation project, which will use both binary cycle equipment and a gas engine powered by the liberated methane, is expected to provide data on the economics of the utilization of geopressured brines. Resource extraction studies are also continuing.

Hot Dry Rock R&D. The development of the science and technology required to produce energy from Hot Dry Rock is proceeding slowly due: 1) to high development costs associated with drilling and inducing artificial fractures, 2) long lead times, and 3) the fact that few experiments have been conducted. Except for areas of high heat flow near tectonic plate boundaries and areas of recent volcanism, the temperatures required are only found at depths of 6 to 7 km. Since drilling down to this depth is expensive, preliminary test projects are either being carried out a lower depth and/or in areas where the thermal gradient is higher than normal, i.e. in areas of geothermal anomalies. Techniques developed in these regions will subsequently have to be tested in typical depths and gradients.

Testing to determine the feasibility of creating large-scale underground heat extraction continues. In July 1986, a major milestone was achieved at the Fenton Hill Hot Dry Rock project. Having connected the two deep wells and creating an underground loop by which water is circulated through a reservoir of fractured Hot Dry Rock, researchers succeeded in extracting water from the reservoir at 190°C. The system appears to become more efficient as it operates, so even higher temperatures are predicted. The next step in this programme, in which both the Japanese and German governments have participated under an IEA agreement, is to prepare for a one-year heat extraction test designed to demonstrate reservoir heat capacity.

Magma R&D. The extreme temperatures and great depth of the resource are problems which must be overcome if magma is to be exploited. New and innovative concepts will have to be developed, and advances in materials and drilling will also have to be made before the resource can be exploited. Preliminary engineering and economic feasibility studies are being funded as well as research to develop equipment and materials capable of withstanding the temperature of the magma environment (up to 1000°C at 3300 to 6500 m.)

Spin-offs from Geothermal Research. Geothermal research and experiments are likely to lead to discoveries beneficial to other fields. The immense underground heat exchangers needed for Hot Dry Rock technology could be used for storage of thermal energy in rock, initially at modest temperatures. These techniques may be of value in stimulating production in gas fields, oil recovery or in situ combustion of coal. Advances in underground exploration may be used in mining, tunneling, seismology and perhaps other fields and as waste disposal. Advances in drilling and downhole instrumentation have obvious spin-off benefits for the oil and gas industry.

Outlook

Geothermal energy currently provides twenty countries with the energy equivalent of more than 60 million barrels of oil per year. Although only a small portion of the total energy supply, this represents enough heat to meet the needs of over 2 million homes in a cold climate and enough electricity for over 1.5 million homes. The past decade has seen a 7-fold increase in geothermal power generation, and if all the projects which are planned become operational world capacity could increase to over 10 000 MW_e by 1995.

The attractiveness of the technology has been enhanced by numerous advances resulting from the R&D programmes of the past decade. Important progress has taken place in exploration methodology and reservoir development technology. Major brine handling and environmental pollution problems have been solved. Liquid-dominated resources can now be exploited economically for electricity production with flash steam plants. If the technology and economics of larger binary plants are improved, enabling utilization of the very large lower temperature resource base, then significant increase in geothermal electric capacity is likely.

The uncertainty of hydrothermal reservoir performance is still a deterrent to more rapid geothermal penetration. The financial community is generally unwilling to consider investment in wellfields and power plants costing over $100 million without reliable quantification of adequate reservoir production over the 30-40 year expected lifetime of the plant. This problem will not be resolved until valid techniques for simulating reservoir performance under production conditions are available and the cost of confirmation drilling is reduced. Another approach is the use of portable units which can be shifted to other fields if a reservoir is exhausted earlier than expected.

Major strides have been made in market penetration of geothermal energy. Probably the greatest geothermal success story to date is electricity production which currently provides 6% of Northern California's electricity at less than half the cost of a new coal or nuclear plant. New steam plants at The Geysers in Northern California and flash steam and binary plants at other sites are likely to raise geothermal energy's contribution to approximately 15% of California's added electric power generating capacity between now and the year 2000. Many other countries are vigorously developing their geothermal resources and anticipate a significant contribution to their electricity demands.

The future market for direct use of hydrothermal fluids is somewhat uncertain and highly dependent on the price of competing energy. Another significant barrier to greater commercialization is the fact that a substantial portion of the geothermal reservoirs are located in remote areas distant from markets, transportation, and workforce. The potential market could be very large, especially for the industrial process heat sector. Approximately 35% of current process heat requirements are at temperatures of 160°C or less, indicating that a substantial portion of the direct heat needs could come from geothermal resources if they are extensively developed and applications are located near identified resource areas.

Despite a few technical problems remaining to be solved and uncertainties about the cost of competing energy, continued development of geothermal resources is likely because of the many advantages of the technology. All of these factors appear to point to the likelihood of continued successful market penetration in the future.

The future course of utilization of energy from geopressured, Hot Dry Rock and magma resources cannot be predicted at this time. However, the potential for economically obtaining even a portion of the high amounts of energy contained in these resources provides ample justification to continue research on the feasibility of utilizing these resources.

VII. MAJOR FINDINGS

There is a very large potential resource base of geothermal energy in several IEA countries. Geothermal resources, where they occur, tend to be very large with the potential of meeting energy needs in an area for many years. Like most energy sources, they are unevenly distributed. There are four types of geothermal resources - hydrothermal, geopressured, Hot Dry Rock, and magma. Hydrothermal technology is the only one presently being commercially utilized, and technologies to use the other resources are still in the research stage.

As a result of over a decade of research and development activity and practical field experience, geothermal hydrothermal technology is now being improved rapidly. System improvements and resulting cost reductions have been achieved for both geothermal electric and direct use technologies. Because of their many advantages, geothermal energy systems are increasingly viewed as having significant potential in many countries.

Wide experience has been acquired over the past ten years in geothermal exploration, drilling and extraction, and important material advances have been achieved which successfully deal with the geothermal environment. Also, production and conversion equipment have been very reliable, and 90-95% availability is not uncommon for either electric or direct use systems.

Effective abatement techniques have been developed to solve many of the environmental problems associated with geothermal technology.

The greatest remaining technical need is for better reservoir characterization techniques.

Recent years have seen impressive gains in geothermal market penetration as technical advances have made the technology even more attractive. Many countries have taken steps to develop their geothermal resources and those having existing plants have installed additional capacity.

Geothermal direct heat is being utilized economically in many countries and for a wide variety of end-use applications.

Geothermal electric plants can provide a steady supply of energy, allowing them to be used as either baseload or peak power plants. Geothermal technology offers the important benefits of flexibility and modularity, which allow additional capacity to be added in comparatively small increments requiring comparatively small capital outlays. The lead-times for installation are relatively short - as little as one year to install a 3-5 MW_e plant and about two years to install a 50 MW_e plant. These characteristics of low-cost flexibility, modularity and short lead-times have special appeal to many utilities today, and for these reasons the utility attitude toward geothermal has generally been a positive one.

Hydrothermal electricity generation (both dry steam and flash conversion) is being used commercially in about fifteen countries, with favorable economics.

Research on large binary cycle plants to enable utilization of moderate temperature fluids are expected to greatly increase the exploitable resources in many countries.

In Hot Dry Rock technology, the feasibility of creating an underground heat exchanger loop has been successfully demonstrated and considerable progress made in understanding the complex processes which take place when rock is fractured at depth. The establishment of the lifetime of reservoirs is the next step in determining the viability of this technology.

Important progress has been made in the technology for utilising geopressured resources. Magma energy is still presently in an early experimental stage.

VIII. BIBLIOGRAPHY

1. Agence française pour la maîtrise de l'énergie, *La géothermie. Une énergie nationale directement utilisable pour le chauffage.* Paris, mai 1983, 32p.

2. Annual Renewable Energy Technology Review, Progress Through 1984, Renewable Energy Institute, 1986.

3. Battista, M.G. and Cataldi, C.: (ENEL), "Production and Utilisation of High Temperature Geothermal Energy. Statistical Data Processing Methodology." Paper presented to the meeting on general energy statistics convened by Conference of European Statisticians in Geneva, September 14th-16th 1981, 21p.

4. Desurmont, M.,"Le coût de l'énergie géothermique". Lecture delivered by M. Desurmont, Head of the Department of Geothermal Energy and Hydroenergy, Bureau de Recherches Géologiques et Minières, Orléans, France, May 1983, 18p.

5. DiPippo, R., *Worldwide Geothermal Power Development 1984 - Overview and Update,* Geothermal Resources Council Bulletin, United States Department of Energy, October 1984.

6. DiPippo, R., *Worldwide Geothermal Power Development, 1985 International Symposium on Geothermal Energy,* edited by Claudia Stone, Geothermal Resources Council, United States Department of Energy, 1985.

7. DiPippo, R., *Geothermal Electric Power, The State of the World - 1985,* Proceedings of the 1985 International Symposium on Geothermal Energy, 1985

8. Economic Commission for Europe, United Nations, Committee on Electric Power, *Seminar on Utilisation of Geothermal Energy for Electric Power Production and Space Heating,* Florence (Italy), May 14th-17th 1984

9. Ente Nazionale per l'Energia Elettrica (ENEL), *Attività nel Settore Geotermica. Programma 1982-1987* and Annexes, Rome, 1982, 30p and 37p.

10. Ente Nazionale per l'Energia Elettrica (ENEL), *Centrali Geotermoelettriche di Larderello e Monte Amiata.* Serie Grandi Impianti, Milano, Italy, 1976, 53p.

11. Ente Nazionale per l'Energia Elettrica (ENEL), *Report on Geothermal Energy* prepared by Dr. Ing. C. Corvi for the International Energy Agency, Rome, April 1983, 14p.

12. Geothermal Progress Monitor, Report No. 9, May 1986, U.S. Department of Energy.

13. Gudmundsson, Jon Steinar, *Direct Uses of Geothermal Energy in 1984,* Proceedings of 1985 International Symposium on Geothermal Energy, 1985.

14. DiPippo, R., *Handbook of Energy Technology and Economics,* John Wiley & Sons, New York, 1983, p. 787.

15. McMullan, J.T. and Strub, A.S.: Achievements of the European Community Second Energy R&D Programme, Commission of the European Communities, General Directorate of Science, Research and development, 1984, 46p.

16. New and Renewable Energy in the United States of America, the U.S. National Paper for the 1981 UN Conference on New and Renewable Sources of Energy, U.S. Department of Energy and U.S. Department of State, June 1981.

17. New Energy Development Organization (NEDO), Geothermal Energy, pp.18-21, Tokyo, Japan. 1984.

18. New Energy Development Organisation (NEDO), *Geothermal Energy Development in Japan,* national response prepared for the International Energy Agency, Tokyo, April 1984, 18p.

19. New Zealand, Ministry of Energy, *Geothermal Energy in New Zealand,* information prepared for the International Energy Agency, Wellington, October 1983, 8p.

20. OECD Report on the Environmental Impact of Renewables Energy Sources and Systems, Paris, 1986.

21. Ungemach, P., *Development of Low Grade Geothermal Resources in the European Community - Present Status - Problem Areas - Future Prospects,* Commission of the European Communities, Brussels, International Conference on Geothermal Energy, Florence, Italy, May 11th-14th, 1982, 41p.

22. United States Department of Energy, Office of Renewable Energy Technologies, Geothermal and Hydropower Technologies Division, *Comprehensive Review of U.S. Geothermal Resources* (for the International Energy Agency) prepared by Meridian Corporation.

23. United States Department of Energy: Hot Dry Rock Geothermal Energy Programme, Documentation provided to the International Energy Agency, April 1985, 31p.

24. U.S. Geothermal Energy Program: Five Year Research Plan, 1986-1990. U.S. Department of Energy, July 23, 1986 (Draft).

25. U.S. Geothermal Technology, Equipment and Services for Worldwide Applications, U.S. Department of Energy, DOE/ID-10130, undated.

CHAPTER SIX: OCEAN ENERGY

A. NATURE OF THE OCEAN ENERGY SOURCE

Ocean energy exists in several forms: the mechanical energy in waves and tidal action and the heat energy absorbed by the ocean's waters. Since the source of energy is slightly different in each case, separate descriptions of these energy sources follow.

I.   WAVE ENERGY

Waves are caused by the interaction of winds with the sea surface; they represent a transfer of energy from the wind to the sea. The energy in a wave is a function of the amount of water displaced from the mean sea level and the orbital velocity of the water particles in the waves. The energy transferred depends on the wind speed, the distance over which it interacts with the water, and the duration of time for which it blows.

Individual waves (Figure 29) can be characterized by their height, distance between crests (wavelength), time between successive crests, and speed. The power in a wave is a function of the rate at which energy is transferred across a one meter line at right angles to the wave direction, and it is expressed in units of kilowatts per meter of wave front. For example, a 150 meter wavelength with a 10 second period between crests and a height of 3 m would have a theoretical power input of 50 kW/m, if such waves occur over a period of time.

Resource Availability

The largest concentration of potential wave energy on earth is located between the 40° and 60° latitudes in both the northern and southern hemispheres, where the winds blow strongest. Because wave energy increases with the distance, or "fetch", over which winds interact with the ocean, those shores located at the end of a long fetch have a potentially large energy source. The United Kingdom, for example, is an excellent site because it is located at the eastern end of long fetches across the Atlantic and in the 50° latitudes. Other favorable areas for the exploitation of wave energy are the eastern coastline of Japan and the western coastlines of Scotland, Norway, and the United States.

FIGURE 29
AN IDEALIZED WAVE

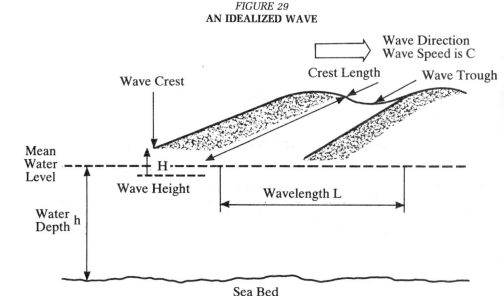

Source: Energy Technology Support Unit (ETSU) Wave Energy, United Kingdom Department of Energy.

II.   TIDAL ENERGY

Tides arise as a result of the gravity of the sun, the moon and the earth's rotation working together. The relative motions of these bodies give rise to different tidal cycles, including a semi-diurnal cycle, a spring-neap cycle, a semi-annual cycle, and other longer cycles. These cycles affect the range of the tides, and knowledge of these variations is important for proper design of tidal power systems.

The amplitudes of these cycles are increased substantially, particularly in estuaries, by local effects such as shelving, funnelling, reflection and resonance. As an example, the combined effect of these factors in the Severn Estuary (United Kingdom) gives rise to one of the largest tidal ranges in the world. The map in Figure 30 shows that the tidal range, which is 4 m at the mouth of the Bristol Channel, is amplified to over 11 m in the vicinity of the Severn Bridge.

The energy of the tides is derived from the kinetic energy of water moving from a higher to a lower elevation. Current designs require a mean tidal range of more than 5 meters before power can be produced. The amount of energy available from the tides is approximately proportional to the square of the tidal range. The energy available for extraction by a tidal power plant

FIGURE 30
TIDAL RESONANCE IN THE SEVERN RIVER ESTUARY (UNITED KINGDOM)

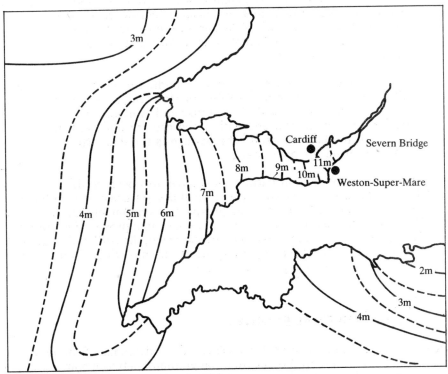

Mean Spring tidal range shown at intervals of 0.5 m

Source: *Tidal Power from the Severn Estuary,* Volume 1; The Severn Barrage Committee.

would vary by a factor of around four over the spring-neap cycle. Moreover, since the time of high water advances each day about an hour, it would not in general be possible to tailor the energy output of a tidal plant to the daily pattern of power demand. However, tides are predictable, and the tidal electricity generating system could therefore be scheduled to make optimum use of available energy.

Resource Availability

During the last fifty years, a number of feasibility studies have been carried out to identify suitable sites for locating tidal energy power stations. The required characteristics of these sites are a large amplitude of the tides together with the possibility of creating large reservoirs to store a large quantity of power-producing water.

Tidal power projects worldwide could theoretically produce 635 000 GWh$_e$. This potential is the equivalent of more than a billion barrels of oil a year. The areas that have the greatest potential are the Bay of Fundy in Canada and the United States; Cook Inlet in Alaska; Chausey in the Bay of Mont St. Michael in France; the Gulf of Mezen in the Soviet Union; the Severn River Estuary in England; the Walcott Inlet in Australia; San José, Argentina; and Asan Bay in South Korea.

The average tidal amplitudes and annual hydraulic energies of several well-suited sites are listed below in Table 38. However, care should be taken in interpreting these figures as the tidal range and available energy will vary with the location of the barrage within the estuary and with the actual construction of the barrage itself. The barrage is the dam structure which divides a basin from the open sea.

A number of potential tidal barrage sites have also been studied in Korea and India. Those in Korea are located on the West coast at Garolim Bay which has a 4.9 m tidal range and prospective annual energy output of 0.8 TWh; and Cheonsu Bay, with a 4.5 m range and a prospective output of 1.2 TWh. In India, two sites have been investigated, the Gulf of Kutach with a 5 m range and 1.6-3.0 TWh potential output and the Gulf of Cambay with a 7 m range and 10-15 TWh potential output.

III. OCEAN THERMAL ENERGY

Ocean thermal energy conversion (OTEC) takes advantage of the naturally occurring temperature difference (thermal gradient) between warm water at the ocean's surface and cold water that is found at depths of about 1 000 m. With the warm surface water acting as a heat source, and the cold deep water acting as a heat sink, a thermal power cycle, or "heat engine" can generate electricity. Such a thermal power cycle resembles a conventional power plant in many ways, except that no fuel is required to warm the water.

Resource Availability

In tropical and sub-tropical areas, the temperature difference between the warm surface ocean water and the 1 000 meter-deep cold water generally exceeds 20°C, which is the minimum difference required for producing energy from ocean thermal temperature gradients. Theoretically, over 20 million square miles of ocean area exist worldwide for possible OTEC sites, potentially supplying tens of thousands of megawatts of electricity.

In general, OTEC plants need to be sited in latitudes within 25° of the equator. The global region of interest thus includes the tropical and subtropical zones, comprising a band extending between latitudes of about 25°S to 32°N. The Gulf of Mexico, the waters surrounding the Caribbean Islands, and territories such as Guam and the Commonwealth of the

TABLE 38

**AVERAGE TIDAL AMPLITUDES AND ANNUAL HYDRAULIC
ENERGIES OF VARIOUS WELL-SUITED SITES**

Location	Average Tidal Range (m)	Hydraulic Energy (10^9 kWh/yr)
Australia		
Kimberley Coast		
Secure Bay I	10.9	2.4
Secure Bay II	10.9	5.4
North America		
Bay of Fundy (U.S.)		
Passamaquoddy	5.5	15.8
Cobscook	5.5	6.3
Bay of Fundy (Canada)		
Annapolis	6.4	6.7
Minas-Cobequid	10.7	175.0
Amherst Point	10.7	2.25
Shepody	9.8	22.1
Cumberland	10.1	14.7
Petitcodiac	10.7	7.0
Memramcook	10.7	5.2
Cook Inlet, Alaska		
Knik Arm	7.5	6.0
Trunagin Arm	7.5	12.5
South America		
Argentina		
San José	5.9	51.5
Europe		
England		
Severn	8.5	14.7
France		
Aber-Benoit	5.2	0.16
Aber-Wrach	5.0	0.05
Arguenon/Lancieux	8.4	3.9
Prenaye	7.4	1.3
La Rance	8.4	3.1
Rotheneuf	8.0	0.14
Mont St. Michel	8.4	85.1
Somme	6.5	4.1
USSR		
Kislaya Inlet	2.4	0.02
Lumbovskii Bay	4.2	2.4
White Sea	5.7	126.0
Mezen Estuary	6.6	12.0

Source: W.H. Bloss and G.H. Bauer, *Renewable Energy Resources*, Report for the World Energy Conference 1980.

Northern Mariana Islands all have excellent possibilities for OTEC sites. In the western Pacific, there are large areas, such as the area surrounding Hawaii, where optimum gradients of 24°C are available between the surface water and the 1 000 meter-deep water. Many of the favorable regions coincide geographically with the locations of developing nations whose ocean thermal resources could provide substantial amounts of locally-produced energy. Figure 31 shows the world ocean thermal gradients and the most favorable locations for OTEC sites.

B. WAVE ENERGY TECHNOLOGIES

I.   INTRODUCTION

Wave power converts the motion of waves into electrical or mechanical energy. In order to harness the energy generated by waves, an energy extraction device is used to drive turbo-generators. Electricity can be generated at sea and transmitted by cable to land, or hydraulic energy generated at sea can be transmitted to a power station on land.

The prospect of extracting usable amounts of energy from the oceans' waves has been investigated for the last 100 years. It was only within the last decade, however, that data on the energy potential of waves has been sought intensively, and many different conversion techniques for electricity generation have been investigated. The United Kingdom and Japan are the most advanced in wave energy R&D.

II.   TECHNOLOGY DESCRIPTION

A wave energy system can be floating or seabed-fixed. The floating type is chain-moored or dolphin-moored offshore. The seabed-fixed type can be either offshore or on-shore and may be built in a breakwater of a port or may also act as a shore protection bank along the coast.

Over the years, numerous wave energy extraction devices have been designed. These devices typically fall into three categories: surface-followers, pressure activated devices, and focusing devices.

—   *Surface-followers* use a mechanical connection between a device that floats on the wave's surface and a fixed pivot to convert the up and down motion of the wave into electricity. One common surface-following device is "Isaac's wave-energy pump." Others include hinges, contouring rafts and nodding ducks. All have in common the fact that a body at the water surface follows the motion of the waves and produces relative motion from which useful energy can be extracted.

FIGURE 31

WORLD OCEAN THERMAL GRADIENT: SURFACE TO 1 000 M (MEAN ANNUAL)

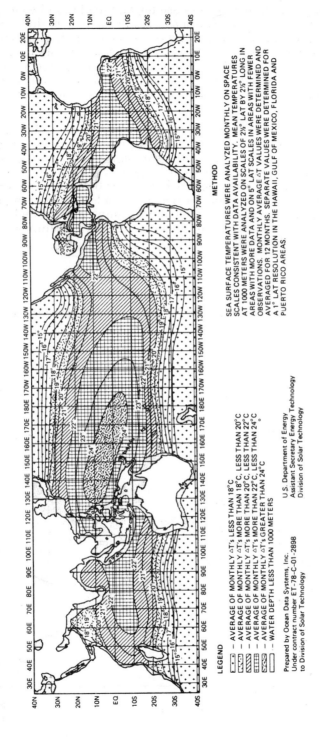

LEGEND

⬚ — AVERAGE OF MONTHLY ΔT's LESS THAN 18°C
▨ — AVERAGE OF MONTHLY ΔT's MORE THAN 18°C, LESS THAN 20°C
▨ — AVERAGE OF MONTHLY ΔT's MORE THAN 20°C, LESS THAN 22°C
▦ — AVERAGE OF MONTHLY ΔT's MORE THAN 22°C, LESS THAN 24°C
▧ — AVERAGE OF MONTHLY ΔT's GREATER THAN 24°C
▧ — WATER DEPTH LESS THAN 1000 METERS

Prepared by Ocean Data Systems, Inc.
Under contract number ET-78-C-01-2898
to Division of Solar Technology

U.S. Department of Energy
Assistant Secretary Energy Technology
Division of Solar Technology

METHOD

SEA SURFACE TEMPERATURES WERE ANALYZED MONTHLY ON SPACE
SCALES CONSISTENT WITH DATA AVAILABILITY. MEAN TEMPERATURES
AT 1000 METERS WERE ANALYZED ON SCALES OF 2½° LAT BY 2½° LONG IN
AREAS WITH MORE DATA AND ON 5° LAT SCALES IN AREAS WITH FEWER
OBSERVATIONS. MONTHLY AVERAGE ΔT VALUES WERE DETERMINED AND
AVERAGED FOR 12 MONTHS. SEPARATE VALUES WERE DETERMINED FOR
A 1° LAT RESOLUTION IN THE HAWAII, GULF OF MEXICO, FLORIDA AND
PUERTO RICO AREAS.

Source: United States Department of Energy, and *Handbook of Energy Technology and Economics.*

— *Pressure-activated devices* use the varying head of water to produce varying pressure. The most successful example of this approach uses the strength of the varying pressure to force air through an air turbine to generate electricity. The oscillating water column concept falls under this category (see Figure 32). Wave motion is converted into the motion of a water column in an air chamber consequently causing an air flow. This air flow is then used to rotate an air turbine to generate power.

Focusing devices use physical barriers to alter the direction and structure of waves to focus their energy to a point, thus condensing the power of the waves. Although these are the simplest wave energy conversion devices, they have a good potential for producing large quantities of energy. One such device is the "wave catching system." Natural and man-made barriers create a large funnel to direct waves into a narrow, steep channel, thus pushing water to higher levels that spill in a catch basin. The water is then released through a turbine to generate electricity.

FIGURE 32
CONCEPTUAL DIAGRAM OF OSCILLATING WATER COLUMN
Type N.E.L. Breakwater

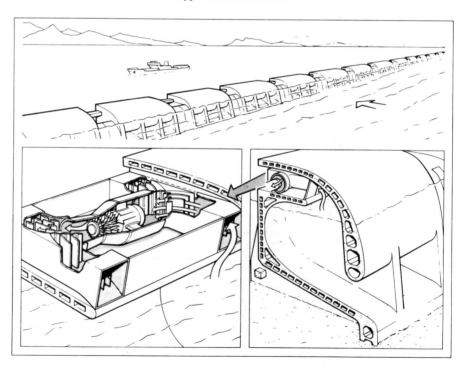

Source: T. Lewis, Wave Energy – Evaluation for C.E.C.

III. **APPLICATIONS**

Wave energy is generally converted into dynamic energy in an air water column or a movable body, and the dynamic energy is, in turn, converted to electricity or hydraulic energy. Wave energy systems have been proposed for baseload power generation: although their daily production varies. However, some concepts have been developed to pump water to land-based reservoirs and baseload generation can be achieved with conventional hydraulic turbines. Wave energy systems may help provide remote power systems for small communities, remote islands and isolated shoreline areas which are not tied into an electricity grid. For example, a wave energy focusing device is being developed for Mauritius, which will have the capability of supplying up to 20% of the island's electricity requirements. An artificial lagoon will be created which not only produces electricity but will also offer favorable conditions for fish farming and mariculture.

In addition, wave energy devices have been developed for power sources for a variety of ocean navigational aids. One example is a common surface-following device employing a simple mechanical system which can cause a bobbing buoy to ring a bell or blow a whistle. More elaborate systems use the up and down movement to compress air in a vertical pipe. The compressed air is released through a turbine which generates electricity to power lights.

IV. **STATUS AND ACCOMPLISHMENTS**

Status in 1974

In 1974, the only significant application of wave energy was the use of small wave power generators for navigational buoys at sea with generating capacities of approximately 0.5 -3 kW. Over 700 of these wave energy generators have been in use around Japan over the last fifteen years, and 500 have been used by other countries.

Technical Advances

Japan. Between 1978 and 1979, Japan's Maritime Safety Agency put wave power devices with a total generating capacity of 500 watts into practical use, seven of which are currently used. A 3 kW wave power generator for lighthouses was designed and studied between 1983 and 1984. In addition, testing of a 120 ton buoy indicated it could operate with a conversion efficiency of 80% and produce 300 kW$_e$ of power.

The first open sea test of a floating wave power generator device was conducted from 1979 to 1980 in the Sea of Japan by the Japan Marine Science and Technology Center. This research barge, known as Kaimei, was tested under the auspices of the IEA by Canada, Ireland, Japan, the United Kingdom, the United States, with Japan acting as Operating Agent. An overall energy conversion efficiency (defined as the ratio of electrical output power to the power of waves having the same width as the barge's length), was approximately 4% when the waves were 3 m high with a period of 8 seconds. Kaimei's hull is 80 m long, 12 m wide, 5.5 m high and weighs approximately 820 tons.

The Kaimei project resulted in a better understanding of floating wave power generators, oscillating water columns, the proper location of air chambers and the location of buoyancy rooms. Moreover, important data on wave climate and wave energy distribution were obtained as well as data on hull movement, mooring system safety and system operation.

Additional IEA experiments with Kaimei began in the fall of 1985 with Japan, Ireland and the United States participating. These completed experiments aimed to improve the generating efficiency, to study and compare different types of air turbines, and to study the use of phase control to increase the energy output.

From 1983 to 1984, the Japan Marine Science and Technology Center conducted open sea tests on a shore seabed fixed-wave power generator test facility called Sanze. This facility was installed on the shore reef of the north-eastern part of the Sea of Japan with its air chamber located in a creek. Various studies were performed on the operation of this 40 kW wave power converter with satisfactory results. It is expected that such a system will be used to supply power to remote islands where diesel power generation is used.

Norway. A Norwegian enterprise has carried out extensive R&D on wave power since 1975. In the 1980s, this R&D culminated in the construction of a 500 kW$_e$ prototype wave power system called Multiresonant Oscillating Water Column (MOWC), which was excavated in a shelf of a cliff on the Norwegian coast. The power generation is achieved by a pressure-activated device which is based on the resonance principle. This design is unique in its ability to maintain a high capture efficiency over a range of periods of the wave fronts, whereas previous designs operate efficiently only at wave front periods that are at, or close to, the resonant frequency of the oscillating water column. The developer estimates that the potential cost of energy generated from this system is US¢5.7/kWh in 1986 prices. Multiresonant oscillating water column plants are now being offered for sale by the manufactuer.

Another private enterprise and the Norwegian Central Institute for Industrial Research developed a novel focusing device for wave energy conversion - the Tapered Channel Wave Power Plant (TAPCHAN) See Figure 33. A pilot plant was built in Bergen, Norway and became operational in March, 1986. The 350 kW$_e$ plant is predicted to generate electricity at a cost of US¢7.2 - 8.6/kWh. The developer plans a commercial small-scale plant to produce electricity at US¢5 - 6/kWh.

FIGURE 33
TAPCHAN DESIGN

Source: Norwave A.S. Tapered Channel Wave Power Plants (Tapchan).

The TAPCHAN power plant consists of a collector, an energy converter, a water reservoir and a conventional hydroelectric power plant. Its unique feature is that the energy converter is entirely passive and almost completely insensitive to variations in wave height and frequency. Moreover, it is based on a concept which is adaptable to a large spectrum of external conditions. The developers have already developed the theoretical basis and adequate design tools for optimization, design and performance analysis of TAP-CHAN power plants which can be sized between 0.5 and 300 MW and are now offering the plants for sale.

Sweden. The use of oscillating buoys as wave energy converters has been extensively studied both in Norway and Sweden. The Swedish Energy Research Commission's R&D efforts have focused on the tube pump converter.

The main feature of the tube pump is the tube itself which is constructed so that it changes volume when it is being stretched. The tube pump converter has been studied both in scale models and in large scale tests. In 1984, the Swedish State Power Authority performed an evaluation of the tube pump converter system which produced energy at a cost of between 0.2 - 0.4 Skr/kWh. Further R&D work is needed before commercialization.

United Kingdom. Ideas for more than 300 devices have been evaluated since the United Kingdom initiated their wave energy programme in 1974. Model devices at scales ranging from one-hundredth to one-sixtieth in size have been tested, and the engineering viability and power outputs have been assessed. Although the United Kingdom's wave energy programme was terminated in 1985, some demonstration projects may be developed with the help of the North of Scotland Hydroelectricity Board through a Commission of the European Communities' grant, and a commercial firm.

Some of the more successful wave energy devices studied under the United Kingdom Programme are described below:

— *The Circular Clam Wave Power System:* The circular clam was originally developed as a linear device using huge bellows which, under wave pressure, forced air through turbines to generate electricity. This pressure-activated device was restructured into a circular version with efficiencies of up to four times that of the original version. Tests indicate that the device could generate power at 2.4 - 4p/kWh, compared to the 6p/kWh predicted cost of the linear device. The prospects of developing a full-scale test are being discussed.

— *The Salter "Duck" Wave Energy Device:* The duck consists of a row of "beaks" that bob up and down with the waves. This surface-following device uses internal hydraulics to capture the energy of the rotating action, and a turbine in the duck generates electricity from the hydraulic pressure. Although initial studies of a $1/10^{th}$ scale model indicated that electricity from the Salter Duck System would cost more than electricity generated by existing plants, newer versions of the system, utilizing a single duck show promise of being significantly cheaper because of easier mooring.

— *The Floating Air-Bag System:* This system uses a ring of air bags to absorb waves coming from all directions. Each wave squeezes the bag it passes, producing high pressure air that is released through a one-way valve to turn a turbine to generate electricity. This pressure-activating device is not as efficient as some, but its simplicity is expected to produce relatively inexpensive energy.

United States. The United States Department of Energy programme's emphasis has been on the analysis, design and construction of the Pneumatic Wave Energy Conversion System (PWECS). PWECS is a pressure-activated device which captures the energy of the up and down motion of waves in a pneumatic chamber by alternatively compressing and expanding air trapped above the moving surface and then passing the air through a uni-directional rotating turbine (see Figure 34). A prototype unit of 125 kW_e was constructed by the Solar Energy Research Institute (SERI). Researchers at SERI and the Johns Hopkins University Applied Physics Laboratory also developed a numerical model of the thermodynamics of the capture chamber and modified the PWECS original design. The result was the unit that was sent to Kaimei for at-sea-testing under the IEA programme.

FIGURE 34
PNEUMATIC WAVE ENERGY CONVERTER (PWEC)

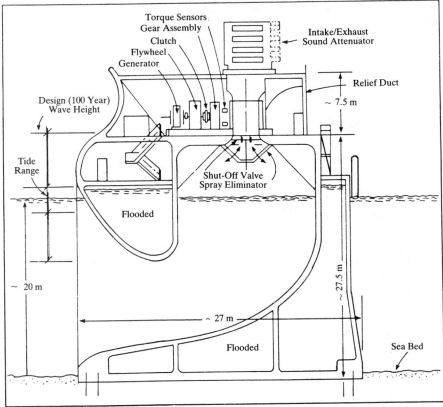

Source: United States Department of Energy.

A dome-shaped, refractive-focussing wave energy device, referred to as a "dam atoll", was also developed in the United States. R&D efforts have resulted in refined dome shapes yielding higher efficiencies and in the development of a model test plan and the design of a 1/50th scale test programme. Studies indicate that a dam atoll device of 100 m in diameter with a central chamber of 11 m in diameter and 20 m deep would produce 4 MW$_e$ from 2 m waves.

In 1982, the United States Government funded a corporation to perform analyses and experimentation on a concept that wave energy can be extracted from sub-surface vertical flaps which are separated by a specified fraction of a wavelength. Initial analyses showed that this "tandem flap" concept could feasibly extract 90% of the wave's energy. A test unit is currently being designed and performance models are being used to determine expected performance and to identify the testing programme needs. It consists of the vertical flaps which are pivoted below the wave surface and responds to the dynamic pressures of the water particle motion. It is expected that the test unit will go into operation (Lake Michigan) in the early summer of 1987 and will run for 3-4 months.

Commercial Status

With the exception of small systems for navigational aids and two small power plants in Norway, wave energy systems have not been commercialized. However, some of the concepts described in the previous section are rapidly approaching commercialization. One United States company is currently proposing to build its own wave energy system, The Neptune, to potential customers, among them the North of Scotland Hydroelectricity Board, the United Kingdom's Central Electricity Generating Board, and the Southwest Electric Supply Board of Australia. The Neptune is a fixed structure device based on the use of concrete caissons. They report that approximately US $3.5 million dollars of private capital has been invested in the system and the company is now seeking backers for a full-size device generating up to 1.2 MW_e from each caisson. In wave energy conditions such as those found off northern Scotland, the company is confident that Neptune could produce power at US ¢3.5 - 5.3/kWh. The company predicts systems could eventually produce electricity in the United States for about US ¢5 - 7/kWh.

A consortium of European companies led by Scottish firms have tentative plans for installing and operating a 5 MW_e breakwater plant off the west coast of Scotland by the summer of 1987. The same team is considering four other possible installations, including a 30 MW_e system in the West Indies and a project for a remote community in Oregon.

Although small-scale floating devices for navigational buoys are economically feasible, wave energy electricity generation has not progressed sufficiently to demonstrate its economic viability. Moreover, wave resources typically have daily and seasonal variations that, unlike OTEC plants, result in significant variations in the wave energy conversion plant output, thus making it nearly impossible to predict the cost of power produced.

A wave energy device's true operating costs can only be determined by tests with systems such as Norway's MOWC and TAPCHAN plants, capable of generating 350 -500 kW. Private financial support is needed for the construction of such plants.

The nearest-term market for wave energy systems is probably for remote power in isolated locations where the technology could be economically competitive with diesel engine generating systems. Nevertheless, further improvement in conversion efficiency, together with efforts for reducing capital investment costs, is required if meaningful market penetration is to occur.

V.   **ENVIRONMENTAL CONSIDERATIONS**

An offshore wave power station would modify the local wave regime which might affect the normal seasonal changes in beaches. This could be a positive or negative impact. Moreover, erosion is created by breakwater energy absorbing devices. A decrease in the wave energy incident upon shores and shallow sub-tidal areas could result in changes in density and species of organisms they support. The residual drift currents would also change, perhaps affecting the spawning of some fish (i.e. herring) whose larvae could be carried away from spawning grounds as well as affecting the migration patterns of surface-swimming fish.

Some offshore wave energy devices would present a navigational hazard because of their low freeboard which would render them relatively invisible either by sight or by radar. Wave energy devices drifting as a result of a mooring failure would also present a danger, not only to ships but to coasts and harbours landward of a station.

Although land-based wave power systems such as the two full-scale prototypes (MOWC and TAPCHAN) built near Bergen, Norway, may have fewer consequences than offshore systems, their potential impact on the environment has not yet been studied.

VI.   TECHNOLOGY OUTLOOK

Current Research Underway

Many different types of wave energy extraction devices have been designed, developed, and extensively studied. Most countries are continuing to test and explore the various concepts developed within their wave energy programmes. More R&D, especially full-scale testing, is needed to improve energy conversion efficiency, to advance system design and integration, and improve components in order to make wave energy conversion economically feasible.

Outlook

Increased production of small-scale wave energy generators (100 W-1 kW) is expected in the future. The shore-fixed type of wave power converter is expected to be put into practical use in the near future as a power source for isolated islands or for various other uses such as waterbreaks. It is expected that the Kaimei plant will probably be put into operation as a multi-purpose unit, serving both as a wave pacifier and an electricity generator.

If wave power generation is to achieve practical and widespread use, it will be necessary to lower its cost through further research and development. It is difficult now to forecast the scale and timing of the potential contribution from wave energy. Research must continue to resolve the remaining technical and economic uncertainties so that both system and energy production costs can be reduced.

VII. MAJOR FINDINGS

— Fundamental research and prototype applications of wave energy technology involving investigation of performance, reliability efficiency, risks, pay-back period and technical uncertainties have been carried out through governmental support (United States) or through both government funding and private industry participation (Japan, Norway, United Kingdom).

— Small wave power generators for navigational buoys at sea and lighthouses are commercially available.

— Open sea tests carried out by the floating wave power generator device, Kaimei, under a joint IEA research project have clarified some of the technical problems that need to be solved prior to commercialization of this type of device.

— The oscillating water columns device, one of the pressure acitivated energy conversion devices, appears to be the most technically advanced from a cost/performance standpoint.

— Some of wave energy utilization's effects on the environment may include impacts on the ecosystem, coastal changes and mariculture. Though thought to be minimal, these potential effects need to be thoroughly investigated and technical solutions developed.

— The nearest term market for wave energy systems is probably remote power systems. Given the immature status of the many wave energy systems, it is too early to estimate the total power production potential.

— The cost-competitiveness of wave power systems remains to be demonstrated. Further improvements in efficiencies are needed. Full-scale sea tests are being performed so that conclusive cost calculations can be made.

C. TIDAL ENERGY TECHNOLOGIES

I. INTRODUCTION

From time immemorial man has sought to use the energy of the tides. The first tidal mills appeared on the coasts of Britanny, Andalusia and England as early as the 12th century. Hundreds of tidal energy plants powered lumber and grist mills in colonial New England.

Tidal electric energy has not yet been commercialized in any sense. The experience to date has been very limited, and the few existing plants are pilot plants; although the La Rance (France) tidal plant is now operating on a commercial basis. The success of these first installations together with the funding of several feasibility studies have given encouraging results as to the cost of electricity generated by tidal power plants. If capital cost and environmental problems can be overcome, several countries will be in a favorable position to exploit their tidal resource.

II.   TECHNOLOGY DESCRIPTION AND APPLICATIONS

Tidal power extracts energy from tides using the same principles as hydropower facilities but capturing tidal ebbs and flows rather than the flow of a river to generate electricity. The simplest systems generate power by damming a cove in an area with high tides. Sluice gates allow rising tides into the cove and block their exit after high tide. The captured water is released through a turbine to generate electricity. More complicated systems impede the flow of water in both directions and generate electricity on the incoming and outgoing tides.

The type of turbine best suited for the low-head characteristics (less than 13 meters) is the axial-flow, bulb-type turbine. The size of the bulb turbines is steadily increasing, and units are manufactured with runner diameters of 7.5 meters, with capacities of 60 MW_e.

The United States Department of Energy has funded design studies on an alternative system that uses a very thin plastic barrier rather than a solid dam. The "water sail" dam is supported by floats. As the system operates, the differential tide height is kept to approximately 2 meters, producing pressures well within the margins of existing plastic membranes. The flow of water from the high side to the low side is used to compress air, which can either be used for the direct generation of electricity or can be stored for later conversion during peak power generation periods.

As an alternative to building dams to tap tidal power, some studies have examined the possibility of siting large turbines in tidal currents to act like underwater windmills. Tidal currents run strongly in and out of some coves and represent a large source of kinetic energy. They would not require dams and would be dispersed over a wide area, thus presenting fewer environmental problems. The technology of such systems is not as far advanced as the hydroelectric turbines used in conventional tidal concepts.

Generation Modes

Single basin tidal power schemes may be designed to operate in one of three different modes: ebb generation, flood generation, or two-way generation.

Ebb generation allows the rising tide to flow in through sluices and turbines, which idle in reverse (Figure 35). Both sluices and turbine passageways are then closed soon after high tide. These are kept closed until the tide has ebbed sufficiently for the difference in water level between the barrage and the sea to drive the turbines and their generators. The water is then allowed to flow through the turbines until the difference in water level is too low to turn them efficiently, when they are closed down. This occurs when the water in the basin is at about the mid-tide level.

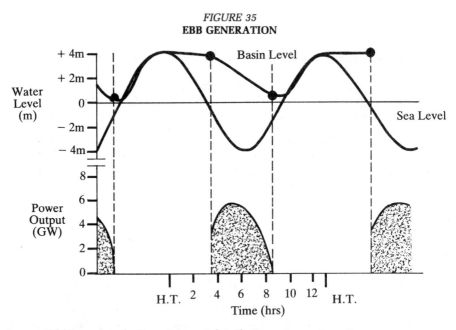

FIGURE 35
EBB GENERATION

Source: *Tidal Power from the Severn Estuary.* Vol. 1; The Severn Barrage Committee.

Flood generation operates in the reverse mode. Water is released from the basin through the sluices until low tide. The sluices are then closed against the incoming tide so that the water level outside the barrage rises above that in the basin. When the appropriate head of water for driving the turbines has been achieved, they commence operation and continue until the water level in the basin has reached about mid-tide. At this time, the head of water across the barrage is no longer adequate to drive the turbines.

In restricting the tidal levels to below mean sea level, a flood generation scheme can have a very severe impact on all ports above the barrage and lead to major visual and ecological impacts. In addition, a flood generation scheme would provide less energy than the equivalent ebb generation scheme. Therefore, this mode of operation should probably be discounted.

Two-way generation combines both modes of operation by generating over parts of both the rising and falling tide. Toward the end of a period of ebb generation, the sluices are opened in order to reduce the basin level quickly to prepare for the period of flood generation. This curtails the period over which ebb generation might otherwise have occurred. A similar situation occurs at the end of flood generation when it is necessary to open the sluices to fill the basin as quickly as possible prior to the ebb operation. This operational principle was applied in the construction of the La Rance tidal power station (France), but the scheme is now operated in the ebb generation mode, augmented with pumping on the flood tide at around high water particularly on neap tides.

The main advantage of two-way generation is that it allows electricity to be produced during a longer period of the day and permits greater operational flexibility. However, this is offset by a number of disadvantages. According to feasibility studies carried out in the United Kingdom for the Severn Estuary scheme, two-way generation would produce slightly less energy than simple ebb generation, would be 15-20% more expensive, and would have a severe impact on ports and navigation.

Pumping. In each operating mode, the turbine could be designed to pump water in the opposite direction to the flow during power generation. In theory, the advantage of regulating fluctuations in supply and demand is an energy gain. Pumping has been applied at La Rance tidal power station. It is therefore possible to super-elevate the basic level with respect to sea level at the end of filling (Figure 36). Power production is thus increased, since the water pumped under a low elevation head (a) will later work under a higher head (b). There can also be a significant gain in energy value from pumping at times when electricity is cheap and generating when the electricity is more valuable.

FIGURE 36
TWO-WAY GENERATION WITH PUMPING

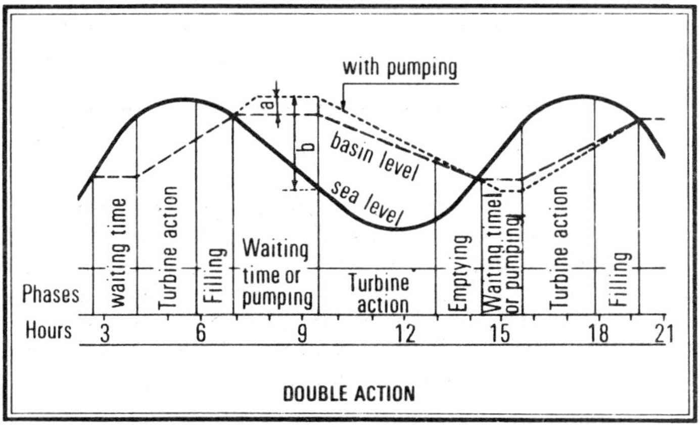

DOUBLE ACTION

Source: The La Rance Tidal Power Station, Electricité de France.

IV. **STATUS AND ACCOMPLISHMENTS**

Plants in Operation. There are very few tidal energy power stations in operation today. The La Rance tidal power station, at the mouth of the La

Rance estuary on the west coast of France, was the world's first tidal plant, and it remains the largest. (See Figure 37) It was conceived as a large prototype for possible larger tidal plants on the Brittany coast. Construction began in 1961 and was completed in 1968. The system uses twenty-four, 10 MW$_e$ bulb type Kaplan turbine generating units to provide an installed capacity of 240 MW$_e$, and a net annual electricity production of about 500 GWhe. Power consumed by pumping represents about 65 GWh/year. The estuary has a tidal range of 14 meters and the 750 meter-long dam creates a 22 km^2 basin, holding 180 million m^3 of useful water.

FIGURE 37
THE LA RANCE TIDAL POWER STATION

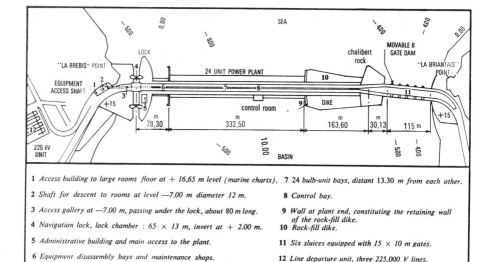

1 *Access building to large rooms floor at + 16.65 m level (marine charts).*

2 *Shaft for descent to rooms at level —7.00 m diameter 12 m.*

3 *Access gallery at —7.00 m, passing under the lock, about 80 m long.*

4 *Navigation lock, lock chamber : 65 × 13 m, invert at + 2.00 m.*

5 *Administrative building and main access to the plant.*

6 *Equipment disassembly bays and maintenance shops.*

7 *24 bulb-unit bays, distant 13.30 m from each other.*

8 *Control bay.*

9 *Wall at plant end, constituting the retaining wall of the rock-fill dike.*

10 *Rock-fill dike.*

11 *Six sluices equipped with 15 × 10 m gates.*

12 *Line departure unit, three 225,000 V lines.*

Source: The La Rance Tidal Power Station, Electricité de France.

The turbines are used to pump water to elevate the generating head. The system could produce a gross generating output of 600 to 750 GWhe; however, it is used primarily for peak power needs. La Rance is highly automated and is operated by only two men on evenings and weekends. A 20 kW$_e$ cathodic protection system has operated quite successfully in preserving the turbines from the corrosive effects of sea water, and the turbine blades were described as "virtually new" in 1984.

The only major fault has been the failure of the lugs holding the generator stators in position in the bulb housings. This has been caused by the severe stress arising during the start-up for pumping. The overall plant availability, excluding the problem with fixing the lugs, has been about 93%.

Another major tidal energy system is the 20 MW Annapolis Tidal Power Project at Annapolis Royal, Nova Scotia, on the Bay of Fundy in Canada, which officially opened in September of 1984. The system was constructed on Hog's Island at the mouth of the Annapolis River at an existing dam built to prevent saltwater from surging onto fertile farm lands during storms. The tidal range is between 4.4 and 8.7 meters. The plant was built to evaluate the feasibility of tidal generation in the Bay of Fundy and of the world's largest straight-flow rim turbine generator (7.5 m in diameter). An innovative powerhouse and intake design, shop pre-assembly of large pieces of turbine-generator equipment, automatic operation and remote controls, as well as other features have been demonstrated at this landmark project.

The cost of the Annapolis Royal plant was approximately US $53 million, or US $2 650/kW (installed). The cost of the generated electricity was projected to be US ¢2.7/kWh. In its first year of operation it has given a good performance with 99% availability. The satisfactory operation of this project has improved the economics of harnessing low-head hydro resources and enhanced prospects for development of major tidal power sites in Canada as well as other parts of the world.

The Kislogubskaya pilot plant, in Kislaya Guba (White Sea, USSR), which utilizes a 400 kW$_e$ generator under a mean tidal range of 3.3 m, began operation in 1967. In China, three small plants of 40 kW, 150 kW, and 320 kW, have been operating over the last several years. In early 1986, China also put into operation a 10 MW tidal power station located at Jianxia in the Province of Zhejiang.

Feasibility Studies. Based on experience at Annapolis Royal, Canadian studies are underway on the feasibility of building a full-scale plant in the Bay of Fundy capable of generating 5 000 MW$_e$ at 3 sites.

One of the studies carried out for Bay of Fundy tidal power concluded that a new approach to design and construction would result in lower unit costs than the previous design. Variable speed turbogenerators were found to be technically feasible and the straight flow turbine was recommended, subject to continued successful experience with the machine of this type now operating at the Annapolis Tidal Generating Station. To minimize powerhouse electrical costs, groups of turbines would operate in internal synchronism, but not necessarily at the frequency of electrical systems. Output of each group would be transformed and rectified for DC transmission.

Many other feasibility studies have been done on the installation of large systems in the Bay of Fundy. In 1979, the United States Army Corps of Engineers reported on approximately ninety alternatives with sizes ranging from 5 to 450 MW$_e$. Projected construction costs ranged from US $22 million to US $916 million.

Another Canadian feasibility study was completed in December 1985 for development of a site in the Cumberland Basin which would link Nova Scotia and New Brunswick and a second site in the Minas Basin. The first site would require 42 variable speed, straight flow turbines for a total installed capacity of 1 428 MW and a new annual energy output of

3 307 GWh at a total cost of US $2.56 billion (January 1984 dollars). The second site would require 128 variable speed, straight flow turbines for a total installed capacity of 5 338 MW and net annual energy output of 13 780 GWh at a total cost of US $7.0 billion (January 1984 dollars). Although having higher unit costs, the Cumberland Basin site was judged to be the more practical and prudent development alternative because of its smaller size, easier financeability due to lower capital costs, and much smaller effects on tidal amplitudes. The study concluded that retimed tidal energy would be competitive in the New England market, while unretimed energy would not. It also suggested that until questions in marketing, retiming and transmission have been resolved, no further preliminary work should take place.

The United Kingdom recently announced that it will invest US $27 million in a three year programme of further detailed studies directed toward building a Severn River tidal power barrage. The United Kingdom Department of Energy will contribute US $6.3 million toward planning the scheme. This follows examination of the results of several studies which began in 1981 on two possible sites on the estuary. The actual plant would cost US $8.25 billion to build, but, according to the study by the Severn Tidal Power Group (STPG) consortium, it could produce power at a cost of US ¢4.5/kWh, compared to US ¢6/kWh for coal and US ¢4.3/kWh for nuclear. Another consortium of seventeen engineering companies and financial institutions has been established to develop a tidal project on the River Mersey.

Commercial Status

With the exception of La Rance, there are no tidal plants operating on a commercial basis. All other existing systems are pilot plants. There is no industry dedicated to building or installing tidal energy systems, but suppliers for all the major components exist. Government-sponsored projects can draw from existing expertise in conventional hydroelectric and marine heavy construction firms. Engineering and construction firms exist with the requisite knowledge and experience to construct tidal energy systems. Turbine design and fabrication services are available from traditional hydroelectric suppliers.

V. ENVIRONMENTAL CONSIDERATIONS

Tidal power installations can have negative impacts on ports and navigation, recreation facilities and various wildlife, particularly birds and fish. Since impacts would be somewhat different for each possible tidal scheme, an assessment should be performed prior to constructing a plant to understand what the likely effects would be. Technical solutions already exist or could be found to solve most of the potentially harmful effects related to a tidal power installation.

Most of the potential impacts arise from three main causes: changes in water levels, water flow patterns and velocities; sediment movement; and the physical presence of a barrage.

Changes in water level influence navigation and the drainage of low-lying land which could lead to major recreational benefits above a barrage. However, this would alter the exposure of the foreshore and inter-tidal banks, thus affecting the habitat of some wading birds and wildfowl.

Changes in water velocity and flow patterns would affect the sediment load carried by the water and sedimentation patterns. These in turn could have important implications for navigation and the natural environment. Water quality in the estuary could be damaged by a reduced dispersion of pollutants because of the presence of a barrage. However, maintenance of the quality of tidal waters would not be technically difficult and could be achieved by introducing more treatment of sewage and industrial waste.

Changes in the pattern of sediment transport, deposition and erosion in the estuary could result in a wide range of possible impacts on ports and the natural environment which require careful investigation since the factors which govern sediment movements are extremely complex. Numerical model studies relating to sediment transport and a physical model to provide detailed information on water movements could identify various effects of a barrage on sediment movements and assist in the adoption of adequate solutions.

VI. TECHNOLOGY OUTLOOK

Technical Prospects

Tidal power is largely based on proven concepts, and there is little need for fundamental research and development. The construction of the major components of a tidal barrage: sluices, turbo-generators, ship-locks and embankments are part of a well-established technology.

So far, most of the feasibility studies on tidal energy deal with very large projects in locations with large tidal ranges. Such tidal ranges can only occur in large estuaries where the resonance effect amplifies the tide, which limits the number of places suitable for tidal energy. These schemes involve huge investments and management problems to coordinate and integrate several very different disciplines. On the other hand, they offer the opportunity of series production both in the civil engineering and the mechanical and electrical work.

Low head turbines were developed for river circumstances (freshwater, constant head) and have been less efficient in varying head situations. However, experience gained with the La Rance tidal station demonstrates that large turbines can be manufactured to work with a high degree of reliability in the marine environment.

Some studies have been made on the possibility of using Darrieus or Musgrove vertical axis turbines to extract kinetic energy from fast flowing tidal streams.

Outlook

The United Kingdom considers tidal energy to be one of its most promising renewable energy sources. The major obstacle is availability of the large private sector investment needed for plant contruction. A proposed 7 200 MW$_e$ plant with an annual output of 14.4 TWh would cost in excess of US $8 billion. However, the plant would be expected to have a life of 100 years, and it is estimated its construction and operation would provide a total of 70 000 jobs.

Tidal power, which is largely based on proven concepts, is capital-intensive and can presently be developed in only a few locations not affected by problems of environmental and social acceptability. These elements limit its possible applications. Although tidal generation does not produce constant power, it is totally predictable.

In any case, since economy of scale is an important factor for the utilization of tidal energy, tidal power projects are likely to be quite large in comparison with other energy sources. Proposed tidal power projects, as in the United Kingdom case and others, estimate required capital investment in excess of a few billion dollars. Very few enterprises in the private sector could carry out such a project even if it is profitable. Therefore, government assistance or participation to mitigate the risks of the huge initial capital investment seem to be indispensable for the development of tidal energy.

VII. MAJOR FINDINGS

Only a few tidal systems are in operation today, the major ones being in France, Canada, and China.

Tidal energy is largely based on proven concepts and technologies; there is little need for fundamental technology development. Tidal energy technology is relatively mature and costs tend to be very site-specific. Therefore, opportunities for cost reductions lie primarily in careful site selection and rigorous control of construction work to limit the high capital costs as much as possible.

The number of suitable sites for development of tidal power plants is quite limited.

Experience gained with the La Rance tidal station demonstrates the technical feasibility of large turbines working with a high degree of reliability in the marine environment. The Canadian installation in the Bay of Fundy has also performed very well.

A tidal power station can affect ports and navigation, recreational facilites, and wildlife because of changes in water level, water flow patterns and velocities, sediment movement and the physical presence of the barrage. Before deciding on the construction of a tidal system, assessment should be made of these potential impacts in order to resolve the possible harmful effects.

The La Rance station in France has demonstrated economic viability for one site, and various studies project promising costs at other sites.

The huge investments required to build large tidal energy plants are a serious impediment to more widespread utilization. Government involvement is probably required to launch a commercial installation.

D. OCEAN THERMAL ENERGY CONVERSION TECHNOLOGIES (OTEC)

I. INTRODUCTION

Ocean thermal energy conversion systems operate on the principle that energy can be extracted from heat sources at different temperatures to produce electricity.

OTEC is one of the few renewable energy options that can serve as a source of baseload electricity, providing power on a continuous basis. It is considered ideal for supplying power to isolated tropical and subtropical areas which have oil-based electrical generation. In addition, the production of fresh water and the cultivation of kelp, shellfish and other marine life using high nutrient cold water can be two valuable by-products of OTEC.

In 1881, the French physicist d'Arsonval was the first to consider the production of electric power through exploitation of the natural vertical thermal gradients in the world's oceans. This idea was developed by one of his pupils, Georges Claude, who conducted experiments over a period of years including the construction of a small shore-based plant off the coast of Cuba, in l930. The plant was technically functional, but did not produce net energy. Although OTEC remained attractive in principle, very little R&D was performed from the 1940s to the 1970s.

II. TECHNOLOGY DESCRIPTION

OTEC plants may be floating or shore-based. There are trade-offs in each approach. Floating plantships require laying undersea power transmission cables. Shore-based plants often involve laying the cold water pipe down steep slopes.

There are two basic power cycles for converting the ocean thermal resource into usable electric power: closed-cycle and open-cycle. In a closed-cycle OTEC power system, the working fluid (typically ammonia or freon) flows through two heat exchangers, a turbo-generator and a boiler feed pump. Warm surface water is pumped through an evaporator, heats the walls, thereby vaporizing the working fluid which drives the turbo-generator. The working fluid vapor flows from the turbo-generator into a condenser, where it is converted back into a liquid by cold seawater transported through a cold water pipe. Figure 38 presents a schematic of a closed-cycle system.

FIGURE 38
SCHEMATIC OF AN OTEC CLOSED-CYCLE SYSTEM

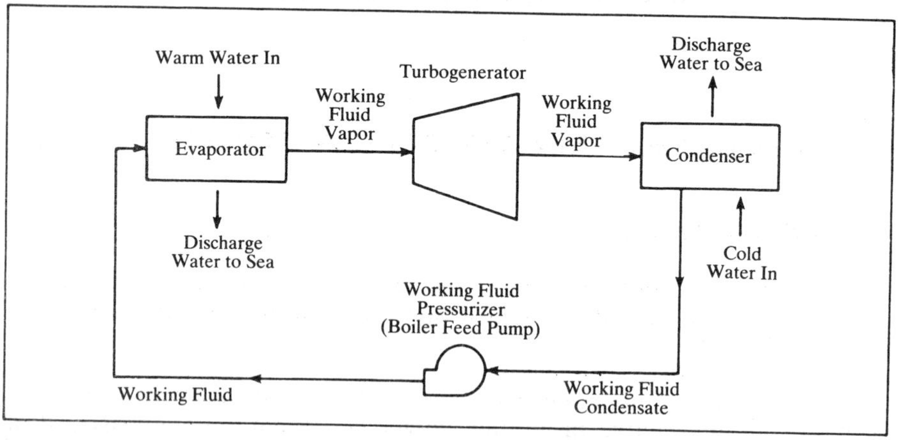

Source: Federal Ocean Energy Technology Program, United States Department of Energy, 1985.

In an open-cycle system, the warm ocean water itself is the working fluid. There are two approaches to the open-cycle system. In the Claude cycle, the warm seawater boils in a vacuum chamber and produces steam which is used to drive a turbine. Cold water from the deep ocean is used to condense the steam and complete the cycle. The other approach is the mist-lift cycle, an advanced open-cycle concept. Warm seawater is flash evaporated at the bottom of a large evacuated lift-tube. The mist then rises to the top of the tube where it is condensed back to a liquid, using the cold deep ocean water. To complete the cycle, electric power is generated when the condensed water flows down a tube, under gravitational pull, discharging through a standard hydraulic turbine. Figure 39 presents schematics of the Claude and mist-lift open-cycle systems.

Closed-cycle systems produce a higher vapor pressure, enabling them to use significantly smaller turbines than the large, low-pressure turbines of open-cycle systems. Moreover, high-pressure ammonia turbines suitable for OTEC applications have been developed in the refrigeration industry and are essentially state-of-the-art. However, cost, corrosion, efficiency, and biofouling problems with heat exchangers have posed significant research challenges.

The open-cycle turbine is somewhat easier to design and build since it does not have to withstand high pressure. But further R&D is needed for large open-cycle plants which have greater power requirements. While substantial interest in the open-cycle concept exists, some researchers believe that closed-cycle systems, which are considered further along in development, have the greater near-term potential for successful OTEC applications. On the other hand, open-cycle systems offer the advantage of fresh water and mariculture by-products (See Applications) which improve their economics.

FIGURE 39
SCHEMATICS OF THE TWO OPEN-CYCLE SYSTEMS:
CLAUDE CYCLE AND THE MIST-LIFT CYCLE

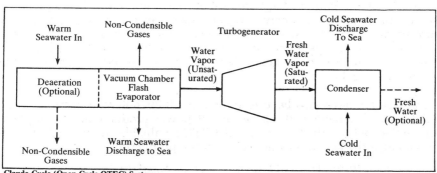

Claude-Cycle (Open-Cycle OTEC) System

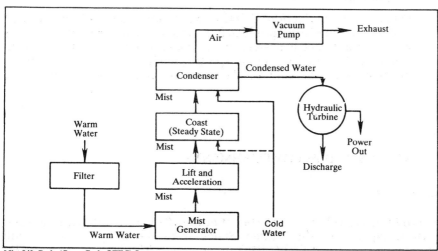

Mist-Lift Cycle (Open-Cycle OTEC) System

III. APPLICATIONS

Baseload/utility-size systems require large (1 MW$_e$ or more) systems that can generate and supply continuous electrical power to a grid. Since ocean temperatures do not change significantly from hour to hour or even month to month, OTEC systems can generate steady levels of power, thus making OTEC particularly suited for baseload electricity generation.

Moreover, OTEC could be ideal for use as remote power systems for small communities, isolated shoreline areas and remote islands that aren't tied into a large electricity grid. Many islands are dependent solely on oil for their electrical power; this factor combined with the transportation costs of the oil make some island sites prime targets for OTEC plants.

Energy for industrial processes could also be derived from OTEC systems for the production of energy intensive, easily-transported substances. For example, floating OTEC plantships could manufacture products such as aluminum, ammonia, hydrogen, chlorine, magnesium, and other sea chemicals which require a large amount of energy to produce. Two potential uses for these energy-intensive products have been proposed. One is that these products could be used as fertilizers, fuels, and feedstocks. The other is that the products generated on the plantship, for example, hydrogen or ammonia, could be transported to shore where they would be reconverted to electricity in fuel cells.

OTEC systems have also been examined for their potential economic and societal benefits beyond the basic use of power production. The applications listed below illustrate some of OTEC's valuable by-products which can enhance the economics of OTEC energy production.

— *Aquaculture.* The cold deep water pumped to the surface for the OTEC plant is nutrient-rich, and can be used to raise a variety of fish or shellfish.

— *Fresh Water Production.* OTEC power plants, especially open-cycle systems, could readily produce fresh water as a by-product. In regions where fresh water is scarce, this application could be of significant benefit, but the logistics of transporting the fresh water probably make the manufacturing economical only for on-shore or near-shore plants.

IV. STATUS AND ACCOMPLISHMENTS

Status in 1974

Modern OTEC development started after the 1973 crude oil price increase. In 1974, Japan and the United States were the only countries with government-sponsored OTEC R&D programmes. The United States OTEC programme, initiated in 1972, first concentrated on evaluating the feasibility of various OTEC options. Initial R&D efforts in Japan focused on a variety of component, material and design studies.

Technical Advances

Although most of the OTEC R&D to date has focused on floating plantships, increased attention has now been given to land-based plants. Moreover, component and material R&D has been a priority among IEA nations. Significant technological advances have been achieved on heat exchangers, power transmission cables and cold water pipes. Corrosion and biofouling have been studied in depth. France, Japan and the U.S. presently lead the world in OTEC R&D. The following paragraphs summarize the efforts made both within and outside the IEA to advance OTEC technology.

Developments in the IEA region

Japan: Japan's Ministry of International Trade and Industry initiated their OTEC programme in 1974 under the auspices of the Sunshine Project. The R&D efforts concentrated on heat exchanger technology, ocean engineering for floating plants, materials for OTEC systems, and design studies for floating and submerged platforms, their associated mooring systems and cold water pipes. Initially, a floating platform system model of 5 kW_e was created to test closed-cycle systems and an oscillating heat exchanger. Tank experiments were performed in order to establish an optimum design method with the highest precision, and a low loss control method. The Japanese Government is presently working toward the establishment of a viable land-based plant, but is also proceeding with designing a 10 MW floating plant to be operational by 1990.

The Tokyo Electric Power Co., Ltd. built a 100 kW_e shore-based closed-cycle pilot plant which it operates in the Republic of Nauru. The plant has a gross power output of 100 kW_e and a net output of 34 kW_e. Based on the favorable results of this experiment, plans are underway for a 20 MW shore-based plant on the same island.

The Netherlands: The Netherlands Government supports some of the efforts made by the private sector in the field of OTEC technology. The Netherlands provided one of the foreign companies contracted to work on a United States Department of Energy study of the technical feasibility of large, 400 MW_e floating OTEC plants. Moreover, a Netherlands firm participated in a feasibility study made by EUROCEAN on a 1 MW_e OTEC plant with desalinization and aquaculture applications.

A feasibility study carried out by consultants from The Netherlands examined the competitiveness of a 10 MW_e floating OTEC plant for the Dutch Antilles (Curaçao or Aruba[1]). Most of the work was on the cold water

---

1. Since January 1, 1986, Aruba is no longer part of the Dutch Antilles.

pipe, mooring and the platform. The researchers concluded that OTEC plants of 10 MW$_e$ could be competitive on small islands, such as those in the Caribbean, where the cost of electricity is high.

In 1982, a project was implemented both by the Netherlands's government and industry to study a 100 kW prototype OTEC plant in Indonesia, at Bali. The effort was postponed in 1985 because the government decided not to finance this costly project at the present time.

Sweden: There is no specific OTEC research programme in Sweden. However, Swedish companies have participated with other countries in various pilot plants. They have been involved in feasibility and design studies, and have demonstrated technological advances in the field of heat exchangers for closed-cycle land-based OTEC systems. The Swedish and Norwegian governments provided support for the design of a 1 MW closed-cycle land-based OTEC demonstration plant in Jamaica. Biofouling and cleaning tests were performed at the test site with different heat exchangers.

United Kingdom: The United Kingdom OTEC Program was initiated in 1981. Technical studies on heat exchangers and cold water pipes were performed in addition to studies on the economic and commercial aspects of OTEC.

In 1982, the United Kingdom Department of Industry and relevant companies began work on the development of a floating 10 MW closed-cycle demonstration plant to be installed in the Caribbean or Pacific. University groups contributed to this project by examining the dynamic response of the system, design of the heat exchangers, and economic and risk factors. The design has been completed, and plans call for installation and operation by the end of 1989.

United States: Initial United States R&D efforts defined the potential feasibility of numerous options. This was followed by baseline designs, computer simulations, laboratory-scale tests, and engineering-scale experimental verifications. Early at-sea testing verified the technical feasibility of some critical aspects of the closed-cycle OTEC configuration. In 1979, a privately-sponsored project known as MINI-OTEC was the first successful at-sea demonstration of a 35 kW floating OTEC plant. An industry consortium and the State of Hawaii mounted this closed-cycle floating plant on a barge and deployed it off the Hawaiian shores. During its four months of operation, MINI-OTEC proved the feasibility of the OTEC concept and produced 15 kW$_e$ net power, the first OTEC plant to generate net power.

OTEC-1, the first large-scale (1 MW$_e$), closed-cycle floating OTEC engineering test facility, was deployed in Hawaiian waters in 1980. Sponsored by the United States Government, the installation provided valuable data on heat exchanger thermohydraulics, environmental impacts, cold water pipe deployment, and the operation and control aspects of OTEC. Due to ensuing budgetary constraints, testing operations had to be curtailed prematurely, and the OTEC-1 project ended in 1981. Figure 40 shows a cutaway view of the OTEC-1 vessel (formerly a U.S. Navy tanker).

FIGURE 40

OTEC-1 FLOATING TEST PLATFORM OF THE UNITED STATES DEPARTMENT OF ENERGY

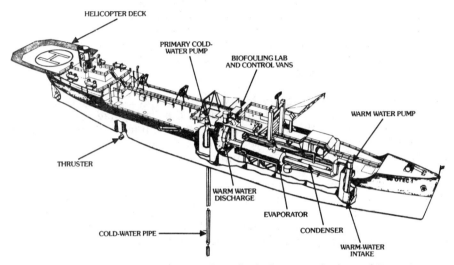

Source: United States Department of Energy, and *Handbook of Energy Technology and Economics*.

Under the United States OTEC Research, Development and Demonstration Act of 1980, plans were made for the deployment of a 40 MW OTEC pilot plant in Hawaiian waters. A closed-cycle near-shore plant was designed featuring a fixed concrete barrier containing the power system equipment. The design effort, headed by a private corporation, was cost-shared with DOE and carried out in association with the state of Hawaii and the Hawaii Electric Company. No further work has been performed due to lack of funding and the decline in oil prices.

The United States research on open-cycle OTEC has led to the definition and experimental evaluation of direct-contact evaporators and condensers with high performance and low fluid dynamic requirements. One United States laboratory is working on a design that combines aspects of both closed- and open-cycle systems called "latent-heat-transfer closed-cycle OTEC". It would flash-evaporate warm sea water under a vacuum and condense the vapor against ammonia which vaporizes to drive a turbine. More R&D is needed before the design can be submitted for testing.

A major advancement in the qualification of lower-cost heat exchanger materials was recently achieved in the United States. Aluminum, originally thought to be too susceptible to erosion, was qualified for long-term OTEC operation through the use of intermittent chlorination. A two-year database resulting from on-site testing indicates no significant biofouling or corrosion in selected aluminum alloys.

France: The French OTEC Program, the most comprehensive in Europe, conducted feasibility studies from 1978 to 1980. After examining both a land-based, closed-cycle OTEC system and a land-based, open-cycle system design, it was concluded that a 5 MW$_e$ land-based pilot plant should be built. R&D activities focused on power cycle component testing, cold water pipe design and testing, long-term biofouling and cleaning tests, and corrosion tests of heat exchanger materials. Tahiti was chosen as the test site, and operation of the prototype plant was expected in 1988. However, problems have been encountered with laying the cold water pipe, and the project may be postponed.

India: The Indian programme focuses on the development of a 1 MW$_e$ closed-cycle pilot plant, to be land-based in the Archipelago of Laccadives. The decision to build this demonstration unit, coupled with a cooling and aquaculture complex, has not yet been made, but studies carried out by private industrial groups are continuing.

Jamaica: Jamaica does not have a special research programme for OTEC but is showing great interest in the development of this technology. The Petroleum Corporation of Jamaica signed cooperative agreements with the Swedish and Norwegian governments to develop a 1 MW$_e$ OTEC unit and to study some technological problems such as biofouling of heat exchangers.

Taiwan: Various studies have been conducted by universities and laboratories, coordinated by the Energy Research Laboratories and Taipower (the national power company). There are plans to build a closed-cycle 9 MW$_e$ OTEC unit which will use the hot water discharged by the future third unit of the nuclear power station of Hon-Tsai. If constructed, this OTEC project will contribute to limiting the environmental impact of the nuclear station.

Commercial Status

No OTEC systems are ready for commercialization at this time. It has been projected that OTEC will not enter the market-place as a viable industry in the 1990s.

V.   ENVIRONMENTAL CONSIDERATIONS

The scientific studies conducted on the possible environmental impacts caused by OTEC have identified several areas of potentially harmful effects. The use of biocides to prevent fouling of evaporators may present pollution problems. The extraction of deep water might cause the release of carbon dioxide into the atmosphere, particularly with open-cycle systems. Coral communities, which are important for ecological and safety reasons, may be adversly affected.

Certain commercial species and life stages of marine biota, such as fish eggs and larvae, may be destroyed if trapped in the OTEC system. This would cause a reduction of commercial species and changes in the local species composition. The impingement and entrainment of biota is a related environmental concern. Changes in the local salinity and temperature might affect the local ecosystem, and have larger effects on the movement of warm ocean currents, temperature, and climate.

Nevertheless, most of the studies have concluded that these effects are limited. For example, a recent Argonne National Laboratory study contends that with low levels of chlorination, biofouling is not a serious problem. Moreover, continuous effort is being made to resolve any potentially harmful effects.

VI. TECHNOLOGY OUTLOOK

Key Work Underway

Commercial development of both closed- and open-cycle OTEC technology requires demonstration of technical performance, reliability and cost-effectiveness. The first step towards meeting these requirements is the construction and successful operation of prototype OTEC plants at sizes of commercial interest, in the range of 1 to 50 MW_e. These plants will not only provide technical performance and cost data, but will also resolve many of OTEC's perceived and actual technological risks.

There are numerous component and material considerations that must be addressed in order to design and construct a commercial OTEC plant that will have the long life necessary to amortize the construction costs. The major components undergoing development are heat exchangers and cold water pipes.

Heat Exchangers: The heat exchangers (evaporator and condenser) are considered to be the most costly and critical items for closed-cycle OTEC systems. Small amounts of biofouling can dramatically affect the heat transfer of heat exchangers. More research is required to develop efficient heat exchangers with adequate cleaning methods to control biofouling. Although material research into cheaper alloys such as aluminum has been performed, additional research is needed to determine their life-span, durability and resistance to corrosion.

Power Transmission Cables. Power cable technology, applicable to the transmission of electric power from a large moored offshore OTEC plant to shore, is not sufficiently developed or tested to assure the long life required of such cables. However, since land-based plants are believed to have the greatest near-term potential, these cables are not currently the focus of major research efforts.

Cold Water Pipe (CWP). Cold water pipe R&D is geared to developing a more efficient and cheaper design for this major cost item, especially for shore-based installations. Various materials have been and will continue to be investigated, including fiberglass reinforced plastic, concrete and steels.

The inaccessibility of the cold water pipe for repair and maintenance poses special research problems. Projections of the service life of the CWP materials need to be made, based on corrosion resistance and solubility studies as well as tests conducted on pressure, temperature and stress conditions.

The technical feasibility of recovery of net electrical energy from thermal gradients in the world's oceans has been proven repeatedly, and a substantial body of literature on the technical, economic, and institutional aspects of OTEC systems has emerged in recent years. The experimental work completed affirms the potential for large-scale utilization of ocean thermal energy both in central power stations and in industrial processes. At present, in the absence of commercial applications, only estimates on market potential are available.

Although the net efficiency of OTEC plants is intrinsically low (because of the limited temperature difference that is available), they are fuel free. Thus, this low net efficiency may be acceptable if the cost of the OTEC-derived energy turns out to be competitive. Although the net efficiency should be as high as possible and must be compatible with system optimization, the key question as to the economic viability of OTEC is the cost of producing usable energy over its life cycle. The major component of the cost of OTEC-devised energy is for the amortization of the substantial capital investment required. The operation and maintenance cost is estimated to be about 1%-2% of the capital investment per year.

According to the United States Department of Energy, the achievement of low operating costs is possible for OTEC because of the absence of fuel costs, the lack of operational hazards requiring human overview, and the simplicity of the system which enables low-cost automation. The achievement of low maintenance costs will be inhibited by the humid, corrosive marine environment but will benefit by the low temperature of operation and the relative absence of high-speed or wear-sensitive machinery.

Since no commercial OTEC plants exist, system costs can only be roughly estimated. The evaporator, condenser, and cold-water pipe (and associated deployment) comprise approximately 70% of the OTEC system cost. Estimates for the capital costs of future technologies for OTEC in the United States are presented in Table 39. Projected sub-system capital costs for both closed- and open-cycle systems are also presented.

Ocean thermal energy conversion technology is slowly emerging from the research stage into the demonstration stage. This needs to be followed by the construction of small commercial plants in the range of 5 to 20 MW. It is expected that OTEC will enter the market in the early 1990s, first providing electricity and/or fresh water, then perhaps energy-intensive products in the long term.

TABLE 39

PRESENT AND PROJECTED OTEC COSTS
(US dollars)

Present Technology (Closed-Cycle)	Interim Goal (Five-Year)	Long-Range Goal
$9680/kW$_e$*	$7200/kW$_e$	$3200/kW$_2$

Sub-System Projected Capital Cost (US $/kW$_e$)

Sub-System**	Present Technology Closed-Cycle	Interim Target (Five-Year)		Long-Range Target	
		Closed-Cycle	Open-Cycle	Closed-Cycle	Open-Cycle
Power System	4120	3200	2900	1500	1280
Building and Structure	560	400	670	290	670
Seawater system	5000	3600	3630	1410	1250

* Dunbar, L.E. "OTEC Market Potential in Small Pacific Islands" International Conference on OTEC Resource Development in the Pacific, Honolulu, Hawaii, October, 1981 ($8400/KW$_e$, adjusted to $5980/KW$_e$ through internal analysis).

** The closed- and open-cycle sub-systems lend themselves to categorization in the shown generic areas. Generally, the components that make up these generic sub-system areas are unique to either the closed-cycle or open-cycle OTEC design and cannot be interchanged.

Source: Federal Ocean Energy Technology Program, U.S. Department of Energy Ocean Energy Multi-year Program Plan, Dec. 1985, page 7.

In the United States, OTEC is expected to first enter into the market in tropical and sub-tropical locations where electricity is produced by oil-fired boilers or diesel-driven generators. Within the 1990-to-2000 time period, OTEC could provide substantial generating capacity for Hawaii, Puerto Rico, the United States Virgin Islands, Guam, Micronesia and American Samoa. A United States study estimated that OTEC could also provide a substantial contribution to the energy demand of the industrialized nations (not including the United States) and the developing countries.

In Japan, it is expected that closed-cycle OTEC electric-power generation will be used by the 1990s to provide power to isolated islands, assuming that satisfactory demonstration plants will have been built and operational by that time.

VII. MAJOR FINDINGS

The potential resource available for OTEC exploitation is, by any standards, very large. OTEC has many applications, from providing base load electricity to producing fresh water. Moreover, it appears that OTEC could provide needed energy for many developing countries.

Interest in developing OTEC prototype plants is widespread. Once a small prototype plant of significant size operates satisfactorily and commercial viability can be projected for a scaled-up size, an OTEC industry could develop.

Intensive R&D efforts must be directed toward improving major components and materials before practical commercial OTEC plants can be built. Of primary importance are heat exchangers, cold water pipes, and research on large (more than 20 MW) open-cycle plants.

Some researchers feel closed-cycle systems have the greatest potential for successful application in near-term commercial OTEC plants. Others feel that in the long run, advanced open-cycle concepts will be the most cost-effective, especially taking into consideration the production of OTEC non-energy by-products which enhance the economics of open-cycle systems.

Deployment of OTEC systems may be potentially harmful to the environment. Among the concerns being studied are the entrainment of biota; prevention of discharges of biocides, corrosion products and working fluids; and affects on local ecosystems. The studies conducted thus far have demonstrated that while these impacts are limited, additional efforts are required to mitigate these possible harmful effects.

VIII. BIBLIOGRAPHY - WAVE

1.   Energy Technology Support Unit (ETSU) for the Department of Energy, *Wave Energy, the Department of Energy's R&D Programme 1974-1983,* document ETSU R26. Harwell, United Kingdom, March 1985, 122 p.

2.   Kvaerner, Brugs A/S, *Multiresonant Oscillating Water Column,* Oslo, Norway, March 1985, 3 p.

3.   Lewis, Tony, *Wave Energy — Evaluation for CEC,* report prepared for the Commission of the European Communities, published by Graham Trotman for the CEC, London, United Kingdom, 137 p..

4.   Mitsui Engineering Shipbuilding Co., Ltd., *The Shore-Fixed Wave-Activated Power Generation System,* Japan, April 1984, 9 p.

5. Moody, George W. and Elliot, George, National Engineering Laboratory, *The Development of the NEL Breakwater Wave Energy Converter,* Glasgow, Scotland, June 1982, 32 p.

6. Moody, George W., National Engineering Laboratory, *The NEL Breakwater - A Realistic Approach to Wave Energy Conversion,* Glasgow, Scotland, October 1983, 21 p.

7. New Energy Develoment Organization (NEDO), *Evaluation of Wave Power,* National response prepared by Japan for the International Energy Agency, Tokyo, April 1984.

8. Norwave A/S, *Tapered Channel Wave Power Plants* (TAPCHAN), Oslo, Norway, October 1983, 10 p.

9. Miyazaki, T. and Ishii, S., Japan Marine Science and Technology Center, *Research and Development on Air Turbine Type Wave Power Generator Systems,* report prepared for the E.C.O.R. International Conference, 1984. First Argentine Ocean Engineering Congress, Buenos Aires, October 1984, Volume I, pp 77-114.

10. Renewable Energy Institute, *Annual Renewable Energy Technology Review, Progress Through 1984,* Washington, D.C., United States, 1986, pp 133-157.

11. Solar Energy Intelligence Report, "Wave Energy Advances Continue Despite Declining Support in U.K." by Mark Newham, June 3, 1986, pp. 174-175.

12. U.S. Department of Energy, *Federal Ocean Energy Technology Program Multi-Year Program Plan FY 85-89,* DOE/CH/10093-100, DE85016870. Washington, D.C., United States, December 1985, 37 p.

BIBLIOGRAPHY - TIDAL

1. Electricité de France, *L'usine maremotrice de La Rance, 15 ans après,* Paris, août 1984, 32 p.

2. Hillairet, P., Vingt ans après, La Rance, une expérience maremotrice (Twenty years after: an experiment in tidal power) in La Houille Blanche, No. 8, 1984, pp. 572-581.

3. Bloss, W.H. and Bauer, G.W., *Survey of Energy Resources, Renewable Energy Resources,* Report prepared for the World Energy Conference 1980, pp. 1273-1352.

4. Boucly, F. and Fuster, S., Energie maremotrice, les conceptions françaises actuelles (Tidal Power, present French designs) in La Houille Blanche, No. 8, 1984, pp. 1597-1605.

5. Electricité de France, *The Rance's Tidal Power Station*, booklet, 17 p.

6. Ministry of Economic Affairs, national response prepared by the Netherlands on tidal energy for the International Energy Agency, The Hague, June 1983, 4 p.

7. Electricité de France, *L'usine maremotrice de La Rance* (The Rance's tidal power station) from the SGE magazine No. 9, November 1984, booklet issued by EDF in January 1985, Paris, 7 p.

8. *Tidal Power from the Severn Estuary*, Volume 1, Report to the Secretary of State for Energy, prepared by the Severn Barrage Committee, Energy Paper No. 46, London, 1981, 112 p.

9. Lewis, J.G., *The tidal power resources of the Kimberleys, The Journal of the Institution of Engineers, Vol. 35. No. 12, Australia, 1963, pp. 333-345.*

10. *Renewable Energy Institute, Annual Renewable Energy Technology Review, Progress Through 1984*, Washington, D.C., United States, 1986, pp. 133-157.

11. Solar Energy Intelligence Report, "U.K. Government To Spend US $8.25M To Study Two Tidal Projects", by Robert McDonald, July 22nd, 1986, p. 234.

12. Solar Energy Intelligence Report, "U.K. To Invest US $27M More in Tidal Power Study", by Robert McDonald, August 5th, 1986, p. 249.

13. Personnel Communication from B. Severn, Engineering and Power Development Consultants, Ltd., August 6th 1986.

BIBLIOGRAPHY - OTEC

1. Bibliographisches Institut Mannheim, *Wie funktioniert das? Die Energie: Erzeugung, Nutzung, Versorgung*, Meyers Lexikonverlag, Germany, 1983, 303 p.

2. Cohen R., "Energy From the Ocean". *Royal Society Conference Proceedings, Marine Technology in the 1990s*, London, 1982, pp. 143-175.

3. Cohen, R., "Ocean Thermal Energy to Feed the Power Grids". *Engineering: Cornell quarterly,* Vol. 17, No. 1, Summer 1982. pp. 12-19.

4. Commission des Communautés Européennes, *Energie Thermique des Mers. Evaluation pour la CEE, Projet de démonstration,* rapport EUR 9895 FR, by Ph. Marchand, CEE, Brussels, 1985, 299 p. (Updated information provided in May 1986).

5. Douglass, R. H., *Handbook of Energy Technology and Economics,* John Wiley and Sons, New York, 1983, p. 877.

6. Lennard, D.E., "Ocean Thermal Energy Conversion - Progress and Prospects". *Proceedings of the Energy Options Conference,* London, 1984. pp. 192-196.

7. New Energy Development Organization (NEDO), *Power From Ocean Thermal Gradients,* National response prepared by Japan for the International Energy Agency, Tokyo, April 1984, 8 p.

8. Ocean Thermal Energy Conversion Systems, Ltd., *Ocean Thermal Energy Conversion - Progress and Prospects* by D.E. Lennard for the Energy Options Conference, London, 1984, pp. 192-197.

9. Renewable Energy Institute, *Annual Renewable Energy Technology Review - Progress through 1984,* Washington, D.C., 1984, pp. 132-157.

10. United States Department of Energy, *Federal Ocean Energy Technology Program,* Multi-Year Program Plan FY 85-89, Washington, D.C., December 1985.

ANNEX II

ECONOMICS OF RENEWABLE ENERGY TECHNOLOGIES

I. INTRODUCTION

Decisions on the development of new energy systems will turn in the first instance on their relative economics. A comparison of costs between systems employing differing primary energy sources is basically a trade-off between capital-intensive projects with low fuel prices and projects with relatively small capital investments and high fuel costs. This is especially the case in the applications of renewable energy technologies (RETs). In general RETs are freer from the future fluctuations in energy prices but are quite capital-intensive compared with conventional energy systems such as those for using fossil fuels. Therefore, capital costs and fuel prices have been the key factors in determining the relative economics of new renewable energy systems.

1.1 OBJECTIVES

This annex to the Review of Renewable Sources of Energy is an attempt to illustrate the economics of renewable energy technologies (RETs) at their current state of the art. This analysis can be used as a point of reference by energy policy makers in understanding the economic benefits and costs of RETs in comparison with those of conventional energy technologies. For each RET, this paper tries to show:

a) the assessment of costs based on, in general, the most advanced or successful applications;

b) an indication of the sensitivity of the above assessment with respect to key factors;

c) future goals envisaged by Member governments whenever data are available.

In addition, estimates of the generation costs of conventional power plants at the point of production and of the fuel costs of conventional heating systems are also provided as a general yardstick for the economics of current RETs.

Because energy policy makers are primarily interested in an optimal allocation of limited resources, efforts were made to remove potential distortions in the analysis. This was done by adjusting for the inherent differences in the ways conventional and renewable energy technologies "deliver" energy to the ultimate users as well as economic differences in such aspects as tax treatment or pricing structures. These adjustments are described in the section on assumptions. It should be noted that these inherent differences mean that this economic analysis should not be extrapolated from use as an "economic yardstick" to use as a prediction of the potential of any of the RETs analysed. Instead this analysis must be combined with the "technical yardstick" provided by Annex I as well as the discussions of key institutional factors in Chapter III of the main body of the report before a more complete assessment of a RET's potential is possible. This is especially true since this economic analysis does not include potential important future impacts on RET economics. Also although illustrations on the sensitivity of RET costs with respect to major factors are shown, they are by no means comprehensive mainly due to lack of reliable data. For instance, sensitivity analysis with respect to economies of scale in production, which is one of the most important factors for the economies of solar technologies in the future, is not included. Sensitivity analyses of other important factors such as major changes in the costs of conventional sources of energy, for example those resulting from oil supply disruptions or increased environmental protection requirements, are likewise excluded.

Due attention should also be paid in the comparison between costs of RETs and those of conventional energy systems. Estimates on the generation costs of conventional power plants at the point of production and the fuel costs of conventional heating systems are no more than indicative. The former is mainly based on the viewpoint of power utilities, while the latter is applicable only for the cases of replacing existing conventional heating systems. The consumers' point of view will in most cases be very different from that of a utility or society. Market penetration is highly dependent on consumer acceptance. Consumers invariably see different prices for energy at the point of use, which *will* reflect the inherent differences in how each energy source is produced, taxed and regulated. Consumers also have very different investment criteria such as discount rates or value of energy security which may become very important in those technologies requiring substantial up-front consumer investment. Again these factors are consciously eliminated from this analysis but must be taken into account when assessing each RETs potential. For these latter purposes in Annex I and to some extent in this annex, whenever data are available, the costs of alternative energy sources for that particular example of RET are quoted.

Finally, although specific RET cost estimates are carried out based on concrete examples, it should be clearly borne in mind that cost estimates are basically site-specific. Therefore, any generalisation from the examples shown in this annex should be interpreted with caution.

1.2 METHODOLOGY

There are at least two standard methods of estimating unit costs of energy:

a) as annual costs per unit of output such as kWh or Btu relating to a share of capital with operating and fuel costs appropriate to a particular year spread over the units sent out in that year;

b) as a levelised cost per unit of output which allocates lifetime production costs over the lifetime output of energy.

The difference between these two methods is whether the time value of money is taken into account or not. The former, in general, does not consider time value, while the latter is a kind of discount cash flow method which adjusts the value of future cash flows by applying appropriate discount rates. To be precise, the levelised cost is the average present value cost in constant money terms per unit of energy produced, at which the total lifetime output of the plant balances exactly the costs of the plant, its operating and fuel costs, plus waste management and ultimate decommissioning. In neither method are the costs of externalities, such as energy security, included.

Annual costs are current costs in one particular year, while the levelised cost is the average over the plant-life in constant US dollars. Therefore the annual cost is useful for finding out the cost borne by the current generation. This is especially the case if prices of energies are linked to the current costs as in the case of electricity tariffs in some IEA countries.

On the other hand, the levelised cost is more appropriate for inter-fuel comparison of systems operating under equivalent conditions because it covers the whole life of the systems under consideration. Since IEA is primarily interested in the optimal allocation of resources from the policy-maker's point of view, IEA has used a *levelised cost method in real terms.*

Another common cost estimation method is estimation of total life-cycle system cost. This method is useful for informing decision-makers about the total investment required for a facility as well as the annual commitment to operating costs. However, the method is relatively cumbersome for use in this type of comparative analysis.

Each of these methods has its uses and exists in several variants which yield different numerical answers depending on the economic or financial ground rules employed and on the technical assumptions considered appropriate. The costs can relate to operating plants, plants under construction or plants under consideration and can be expressed in current, constant or discounted money.

Costs included. A key factor is the point at which a cost estimate is made. Ideally cost should be estimated at the point of consumption by the end user. In the case of electricity, costs to the end users are the system costs which include generating, transmission and distribution, administration and overhead costs. However, for inter-fuel comparison, the allocation of the costs of transmission and distribution and overhead to each energy source and to each type of end user is very difficult and could even be arbitrary. Therefore, in the case of electricity the analyses below use the cost at the point of production (busbar). On the other hand, in the cases of technologies providing thermal output direct to the end-user, it is not necessary to specify the source of energy such as coal, oil, gas and/or electricity displaced and to include costs to the end-user. Therefore costs at the end-user point are exhibited in these cases in this analysis. The units of cost measurement presented in the cost estimates are expressed in the local currency where the example is located, together with the cost of alternative energy for that specific example. Cost estimates are shown in 1984 prices in many cases; although where this is not possible readers are provided with information on the costs of alternatives in the same year.

Discount rate. As shown in Table 40, many IEA countries utilize a real discount rate of 5% or below for the evaluation of their projects. Therefore, in the following analyses, a real discount rate of 5% is assumed. Constant money discount rates in use today vary from country to country, reflecting the considerable disparities in the capital markets, sources of financing, opportunities for investment and a different perception of risk and uncertainty. Private investors generally decide the appropriate discount rate in the light of their own circumstances in dealing with their own cases and these discount rates will in most cases be very different than governments' discount rates.

Subsidies such as preferential tax treatments are excluded in this analysis to the maximum extent possible. In this way a clearer yardstick results, and policy makers can then evaluate the need for subsidies if acceleration of a RETs progress toward economic status is necessary and desired.

Renewable Energy System Costs. Comparisons of costs of energy produced from different fuels require a number of assumptions about key determinants. Each cost factor and assumption is inevitably site specific, and therefore calculated costs must be treated only as indicative. In practice, the situation may differ from one country or even from one site to another. After all, as shown in the conceptual figure below, costs of any energy system could fall anywhere in a wide range depending upon specific conditions of each application. The width of the range illustrates the influence of both risks and maturity on the cost of the technology. The cost estimates of RETs used in this Annex are based mainly on successful cases which would fall in the shaded area B in Figure 41. In the analyses of RET cost estimates below, the costs of the alternative energy systems at those particular sites are presented whenever these figures are available. Using costs in this range will tend to offset at least in the near-term, economies of scale which are not reflected in the analysis.

Conventional Electricity Generation Systems Costs. The costs of conventional energy systems that were used in this analyses are provided as a reference to the current economic status of RETs. It should be noted that, even in the cases of cost estimates of conventional energy systems, there is no uniquely correct figure. For instance, the cost of electricity differs substantially from one consumer to another reflecting differences in their site-specific conditions and forms of demand. The cost estimates of conventional energy systems used in this analysis represent the mode or point A in Figure 41. In short the indicative average cost estimates of conventional energy systems shown below should be considered only to be rough references. Because oil generation resulted in the highest conventional energy system cost, oil generation costs are used as the specific reference in the discussions that follow on each RET.

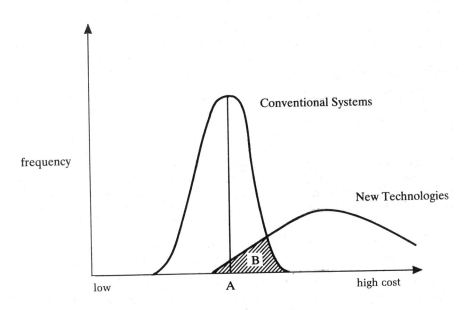

FIGURE 41
DIFFERENCE IN MATURED AND IMMATURE TECHNOLOGIES

Generation Costs of Plants Based on Oil, Nuclear and Coal. Indicative generation costs of new oil, nuclear and coal-based power plants commissioned for 1990 are shown in Table 40 in 1984 currency values. In each case, the most economic plant sizes are chosen. All plants considered are assumed to operate at a 70% capacity factor and to have an economic life

TABLE 40
NATIONAL DISCOUNT RATES
IN MAJOR OECD COUNTRIES
(%)

Belgium	8.6
Canada, Ontario Hydro	4.5
New Brunswick	6.6
Finland	5
Germany	4-4.5
Italy	5
Netherlands	4
Norway	7
Portugal	9-14
Spain	5-8
Switzerland	6
Turkey	5
United Kingdom	5
United States	5

Source: "Projected Costs of Generating Electricity from Nuclear and Coal-fired Power Stations for Commissioning in 1995", NEA/OECD, Paris, 1986.

of thirty years. The capacity factor is chosen considering the likely operation of power plants as baseload. Technically, the capacity factor is often higher depending on the type of fuel used for generation. The table compares the relative economics assuming the construction lead time of three years for oil, six years for nuclear and four years for coal. In the case of nuclear power, a plant with longer construction lead time (ten years) is also examined.

As for coal prices, two cases are examined so as to reflect the situations both in coal-importing and coal-producing regions. For coal-importing regions, coal prices are assumed at US $55 per tonne, while US $38 per tonne is assumed for coal-producing regions. Steam coal prices for power generation in coal-importing regions in the second quarter of 1986 were actually 30%-50% lower than the US $55 per tonne assumed for this analysis. Provided coal prices in these regions stay at, on the average, around US $40 per tonne, the coal price assumption of US $55 per tonne implies an approximately 2.7% per annum. increase over the life of the power plant. The calorific value of coal is assumed to be 6 500 kcal per kilogramme.

Fossil fuel plants capital costs include the cost of FGD and other pollution control measures including combustion modifications for NO_x.

Prices of high sulphur heavy fuel oil are assumed to be US $180 per tonne, or US $4.5 per million Btu, the prices prevailing in North America and Europe in 1984. The assumed heavy fuel oil price of US $180 per tonne, which roughly corresponds to the crude oil prices of US $26-28 per barrel, is approximately 70%-90% higher than oil prices in May 1986. Therefore the heavy fuel oil price of US $180 per tonne means that real oil prices are assumed to increase by 4%-5% per year over the useful life of the plant. This is in line with the 4.6% actual annual increase of oil product prices in OECD countries over the 1978-1985 period.

As shown in Table 41, given these fuel price assumptions, the generating cost of oil is 55.6 mills/kWh, while nuclear and coal cost 29.2-30.8 mills/kWh and 30.7-36.4 mills/kWh respectively. Since uncertainty as to future oil prices is quite high, the generating cost estimate of oil-fired power plants as a function of oil prices are also shown in Figure 42.

Quite naturally, a different set of assumptions bring about different results. For instance, under a 10% discount rate, the generating cost of oil is 61.0 mills/kWh, while nuclear and coal cost 42.2-47.9 mills/kWh and 38.6-44.3 mills/kWh. In the rest of this study, for the purposes of

TABLE 41
GENERATING COSTS ESTIMATES
(1984 US mills/kWh)

| | Oil with FGD* 2 x 600 MW | Nuclear PWR 2 x 1100 MW | | Coal with FGD** 2 x 600 MW | |
		Lead Time		Coal Importing Region	Low Coal Price Region
		6 years	10 years	US $55/tonne constant	US $38/tonne constant
Capital costs	8.2	15.2	16.8	11.1	11.1
O & M costs***	4.2	5.0	5.0	5.0	5.0
Fuel costs	43.2	9.0	9.0	20.3	14.6
Total cost	55.6	29.2	30.8	36.4	30.7
Reference:					
Capacity factor	70%	70%	70%	70%	70%
Construction leadtime	3 years	6 years	10 years	4 years	4 years
Capital investment (US $/kW)	807	1505	1659	1103	1103
(Initial investment) (US $/kW)	(750)	(1300)	(1300)	(1000)	(1000)
(Interest during construction US $/kW	(57)	(205)	(359)	(103)	(103)
Fuel cost	US $180/t				specified above
Conversion efficiency (net)	36%	34%	34%	36%	34%****
Heat rate (kcal/kWh)	2400	2500	2500	2400	2500

(*)   Plant with Flue Gas Desulpherisation (FGD) using high sulphur oil.
(**)   Besides SO$_x$, coal produces various pollutants such as NO$_x$, dust, particulates and ashes. The costs of removing these pollutants to meet legal requirements are included in each cost component.
(***) Operating cost in this study is the cost directly incurred in a plant. Actual operating cost could be higher with the distributable costs of overhead expenses, which differ by utility.
(****) Inland location, lower conversion efficiency.

Source: "Electricity in IEA Countries", IEA/Paris, June 1985, p. 195.

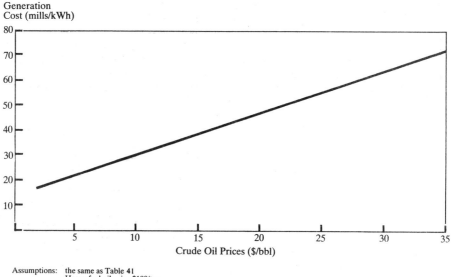

Assumptions: the same as Table 41
Heavy fuel oil price $180/ton
= Crude oil price $26/bbl

comparison with RET electricity generation, when a RET replaces only the fuel costs of conventional generation, the comparative cost used will be US ¢4-5/kWh (40-50 mills/kWh). When it displaces capacity, O and M, and fuel costs, the comparative cost used will be US ¢5-6/kWh (50 to 60 mills/kWh).

Again, it should be noted that comparison between these generating costs and costs of RETs shown in the following section should be done very carefully. Besides uncertainties, which are inherent in these estimates, it should be noted that these are not end-user prices but costs at the point of production or at the site of power plants. In many cases, RET facilities are installed at the sites of end-users and, therefore, end-user electricity prices should be compared with RET costs. Direct comparison between RETs and conventional systems based on the estimates in Table 41 can be made only in the case of power companies which consider RETs as options for future capacity expansion and for reducing operating costs.

Fuel Cost of Heat. In some applications of RETs, such as active solar hot water systems or geothermal district heating systems, the output of the system is heat rather than electricity. For these technologies, the levelised cost estimates must be compared in terms of MBtu delivered. As a reference for these, the fuel costs of heat by various energy sources at the end-user levels are presented in Table 42. The table assumes conversion efficiencies of 90% for electricity and 75% for other fuels (oil, coal and gas). In practice efficiencies differ substantially from one system to another. Therefore the figures in Table 42 are subject to even more stringent qualifications than the previous cases of electricity.

The heat costs shown in Table 42 should be the cost reduction target for RETs when they supplement existing conventional energy systems. However, in the case of new independent applications, the total costs, which include O & M and capital costs in addition to heat costs shown in the table, should be compared. In this case, the target for RET cost reduction would be higher than these heat costs. Unfortunately the O & M and capital costs of each system vary widely, and no generalisation is possible.

TABLE 42
HEAT COSTS IN MAJOR IEA COUNTRIES IN 1984
(US $/MBtu)

	Canada	Germany	Italy	Japan	U.K.	U.S.
1. Industry						
Light Fuel Oil	10.6	9.2	11.8	13.9	8.7	9.0
Electricity	n.a	15.3	n.a	29.6	11.4	16.6
Gas	3.6	n.a.	6.3	14.8	4.5	5.3
Coal	2.5	4.5	2.3	2.6	3.3	2.1
2. Households						
Light Fuel Oil	10.5	10.5	13.6	15.7	10.4	11.1
Electricity	n.a.	27.3	24.7	40.4	22.5	24.4
Gas	5.6	n.a.	9.3	21.4	8.1	7.8
Coal	n.a.	11.8	8.3	n.a.	6.1	n.a.

Conversion efficiencies are assumed at 90% for electricity and 75% for other fuels.

Source: IEA Secretariat estimate based on "Energy Prices and Taxes - Fourth Quarter 1984" IEA/OECD. Calorific values not specified in the above source are assumed to be 6 500 kcal/kg for coal and 10 000 kcal/kg for oil.

II. Analysis of Levelised Costs of Renewable Energy Technologies

A comprehensive economic analysis of all renewable energy technologies is difficult due to the large number of different systems in operation, the wide range of possible applications, the variation in resource intensity and the lack of adequate data on cost and performance.

The following discussion of the economics of renewable energy technologies will focus mainly on factors affecting their economic viability with the use of examples to illustrate actual applications of individual renewable technologies. The technologies are discussed more or less in order of those that are the most economically viable to those that need the most R&D effort to become economic. Future technologies are not analysed. Principally the examples in this chapter represent the most advanced cases. In the actual applications of RETs, the costs of these technologies could be in the wide ranges. Conditions from one project to another can differ substantially, and due consideration should be given to the differences.

1.   *Wind*

Although large WEC systems require additional research and development, small- and medium-size WEC systems have already been commercialized in some OECD regions, particularly in the United States (California).

Figure 43 shows an example of the economics of a proposed 6.8 MW multi-unit WEC installation by a private partnership which is planning to sell electricity to Pacific Gas and Electric (PG&E). The capital investment of US $1 000/kW represents current market prices[1]. Operation and maintenance costs are assumed at US ¢1.5-2.5/kWh based on a current study by the United States Electric Power Research Institute (EPRI)[2]. Uncertainty as to the reliability of WECS is still high under the current state of the art. Although the base case of the example assumes an availability of 90%, in general, EPRI assesses that the availability of small WECS could range from 70% to 96%. Therefore Figure 43 shows the generating costs by availability. In addition, since long useful life has not been demonstrated yet, the figure also presents estimates based on the useful lives of twenty and thirty years.

Under most favourable circumstances (30-year useful life, O&M costs US ¢1.5/kWh and 96% availability), the generating cost of the system is as low as US ¢4.48/kWh or well below the planned selling price of US ¢7.3/kWh to PG&E. Even under the base case assumption, it could be approximately 20% less costly than the PG&E's purchasing price if the system could operate for thirty years. Even if the availability is as low as 70% (low case), the generating cost of the system could be only marginally higher than the planned PG&E purchasing price of US ¢7.3/kWh. Moreover compared with the IEA's estimate of generating cost of new oil-fired power plant of (US ¢5-6/kWh), the generating cost of the wind system could be less costly. In general, under favourable circumstances, a small/medium size WECS is fully competitive with the conventional systems.

With regard to the future possibilities of cost reduction, small/medium WECS have a fairly high probability of becoming competitive with conventional energy systems in the coming ten years even under less favourable conditions than the case in California. Careful assessments of sites are essential. This is particularly the case because the energy content of

---

1. According to "Developments in Exotic Power" (Power Engineering/May 1984), some machines in the market are advertised at prices of US $1000/kW'.
2. "R&D Status Report - Wind Turbine Operating Experience and Trends", Darwin Spencer, EPRI Journal, November 1984.

the wind is proportional to the cube of the wind speed[1]. "Thus at a site with a wind speed of 3 m/s, the energy content of the wind will be 6 000 kWh, whereas a site with a wind speed of 6 m/s will have an energy content of about 48 000 kWh[2]." This implies that the generating cost of a WECS can be reduced by more than 80% if the wind speed is doubled.

Capital investment per kW has already been reduced substantially in the past decade and can be reduced below US $1 000/kW[3]. In fact, capital investment of less than US $1 000/kW is quoted by a WECS manufacturer in Denmark[3].

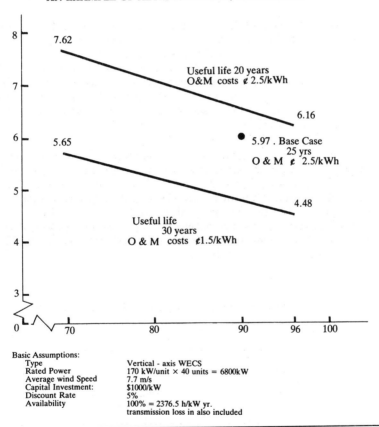

FIGURE 43
CURRENT WECS GENERATING COST BY AVAILABILITY
AN EXAMPLE OF ALTAMONT PASS, CALIFORNIA

Basic Assumptions:

Type	Vertical - axis WECS
Rated Power	170 kW/unit × 40 units = 6800kW
Average wind Speed	7.7 m/s
Capital Investment:	$1000/kW
Discount Rate	5%
Availability	100% = 2376.5 h/kW yr.
	transmission loss in also included

1. "Handbook of Energy Technology and Economics - Chapter 19, Wind Power" by Dennis G. Shepherd, John Wiley and Sons, New York, 1983, p. 826.
2. "Helpful Hints on Windmills in Denmark". Renewable Energy Information Service Technological Institute, September 1984.
3. Estimates provided by VESTAS in Denmark quote DKr. 8545/kW (US $ 776/kW, 11DKr. = 1 US $) for its 55kW wind turbine as of December 1984.

Many experts predict a substantial improvement in the economics of small- and medium-size WECS in the coming several years if certain amounts of production could be assured. However, it is difficult to estimate how great these economies-of-scale might be.

Further improvements are expected in two other fields making small/medium WECS even more competitive with conventional systems in more instances.

First, the availability of WECS can be improved. According to the estimates by EPRI, availability can be improved from the present levels of 70%-96% to 95%-98%. If this were achieved, WECS could be a much more reliable and predictable source of energy than it is today. A 10 percentage points gain of a WEC system's availability reduces its generating cost by about US ¢1/kWh.

Second, given the substantial reduction in capital investment mentioned above, efforts to reduce O & M costs would become more important than before. The above study by EPRI also predicts that O & M costs can be reduced more than one half from the present US ¢1.5-2.5/kWh to US ¢0.7-1.0/kWh in the case of small systems. Since the small/medium WECS is already fully competitive with conventional energy systems, a US ¢1/kWh decline in O & M costs can be of vital importance for the future penetration of the WECS.

Even without projecting further improvements in capital investment (below US $1 000/kW), if improved O & M costs (US ¢0.7/kWh) and availability (98%) were realised, the generating cost of small/medium WECS under the wind conditions like Altamont Pass, CA could be below US ¢4/kWh given the system useful life of 30 years (Table 43). Since the fuel cost of oil-fired power stations was in the range between US ¢4-5/kWh in 1984, WECS could be fully competitive even with existing oil-fired stations under the level of oil prices which prevailed in 1984.

TABLE 43
FUTURE PROSPECTS OF SMALL/MEDIUM WECS

	Current Status	Future Status
Capital Investment (US $/kW)	1000	1000
O&M Costs (US ¢/kWh)(1)	1.5-2.5	0.7-1.0
Availability (%)(1)	70-96	95-98
Generating Cost (US ¢/kWh)(2)	4.48(3)-7.62(4)	3.5(5)-4.2(6)

(1) EPRI Journal, EPRI November 1984
(2) IEA Secretariat estimates based on a 5% discount rate and wind conditions at the site of Flowind Partners II.
(3) 30-year levelised cost with O&M costs of US ¢1.5/kWh and a 96% availability.
(4) 20-year levelised cost with O&M costs of US ¢2.5/kWh and a 70% availability.
(5) 30-year levelised cost with O&M costs of US ¢0.7/kWh and a 98% availability.
(6) 20-year levelised cost with O&M costs of US ¢0.7/kWh and a 98% availability.

Uncertainty is still high in the case of large megawatt class systems. Pacific Gas and Electric Company's MOD-2 turbine costs US $4 000 per kW[1]. Using the same set of assumptions as the case of the smaller systems above, this corresponds with the generation cost of US ¢20/kWh, which exceeds that of new oil-fired power plants by a factor of three to six.

FIGURE 44

CURRENT GEOTHERMAL GENERATING COSTS BY USEFUL LIFE

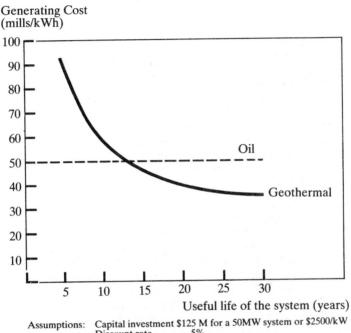

Assumptions: Capital investment $125 M for a 50MW system or $2500/kW
Discount rate   5%
Capacity factor   80%

Source: IEA estimates based on data provided by Japan for this review.

2.   Geothermal

Although the costs of utilization of energy from geopressured, Hot Dry Rock and magma resources cannot be predicted at this time, substantial experience has been already accumulated in the fields of hydrothermal electric and direct heat applications. As early as in 1976, Pacific Gas and Electric reported that their dry steam geothermal power plants produced electricity at the lowest cost of any type of steam plant in its system. The figures in 1982 prices are as follows :

Geothermal   29 mills/kWh
Nuclear   39 mills/kWh
Coal   42 mills/kWh
Oil-fired   58 mills/kWh

1. "PG and E's Evaluation of Wind Generation" by D. Smith. Presentation at the 5th ASME Wind Energy Symposium, New Orleans, February, 1986.

With regard to direct heat applications, although a wide range of costs has been observed, some cases are competitive with oil. In the United States, a study of twenty geothermal direct heat systems shows a wide range of costs from US $1.34 to US $33.94 per MBtu in 1979 prices, while the average price for heating oil sold to residential customers at that time was US $4.75 per MBtu. Eight of the twenty cases above exhibit costs lower than US $4.75 per MBtu of heating oil[1]. In France, the average cost of geothermal energy for direct heat systems based on a representative sample of thirty installations in service or in the course of completion was the same as heavy oil number two in January 1983[2].

According to an estimate by the New Energy Development Organization in Japan, capital investment for a new 50 MW flashed steam geothermal power plant could amount to 30 billion yen or US $125 million (240 yen = US $1)[3]. Figure 44 shows the generating costs of this geothermal power plant by its useful life assuming a capacity factor of 80%, 5% discount rate and O&M costs of 12 mills/kWh. As shown in the figure, geothermal could be fully competitive with oil-fired power plants which cost 50-60 mills/kWh if the useful life of the geothermal plant is over ten years. The 80% capacity factor is commonly achieved in Japan. Although no precise data on O&M costs is available, 12 mills/kWh could not be grossly out of line with the actual O&M costs[4].

As to future prospects, Annex I pointed out that the major remaining technical uncertainty surrounding the use of hydrothermal resources centres primarily on the prediction of reservoir performance. The availability of financing is limited by the perception of risk associated with unpredictable reservoir performance. Therefore better reservoir definition technologies to increase the capability of predicting reservoir productivity and longevity along with fracture mapping technologies for optimal siting of wells are being developed. The remoteness of many geothermal reservoirs is another limit to increased use of geothermal resources, especially for direct heat applications. Location of new industrial process heat users near geothermal resource sites could increase utilization. No quantitative predictions on the likelihood of this occurring were available for use in this analysis.

A considerably high risk is therefore still involved in the development of geothermal power plants and heating systems mainly due to difficulties in the assessment of reservoir performances. However, experience of the past decade in the United States indicates that risk can be limited to about 10% of the total projected system cost, through judicious reservoir testing prior to committing to power plant construction and use of modular well-head generating plants[5]. Geothermal power in these applications has already

---

1. Data provided by the United States for this review.
2. "Le Coût de l'énergie géothermique", lecture delivered by Mr. Desurmont, Head of the Department of Geothermal Energy and Hydroenergy, Bureau de Recherches Geologiques et Minières, May 1983.
3. Data provided by Japan for this review.
4. According to the illustration of "Geothermal Energy" (vol. 8, No. 1, 1983), the O&M costs of a geothermal power plant in Japan is 1.92 yen/kWh (8 mills/kWh. 240 yen = US $1) in 1982 price.
5. Data provided by the United States for this review.

firmly established its economic viability in many parts of the IEA once the reservoir is successfully developed with a proper scale. The binary cycle generation, which is a new process using hot water, is also expected to be competitive with traditional steam-flashed geothermal power plants both in the United States and Japan.

3.   *Biomass*

There are a number of variations in biomass systems depending on the combination of the feedstock, the conversion technology and the form of energy product. As shown in Table 44, feedstocks can be classified into three broad types; organic wastes, land and water-based biomass. Besides wood grown specifically as an energy crop, the most popular feedstocks include straw, wood waste from timber processing, logs and forestry thinnings, chicken litter, municipal and industrial wastes. As for conversion technologies, major processes are:

(1)   direct combustion;

(2)   anaerobic digestion;

(3)   pyrolysis;

TABLE 44
SUMMARY OF FEEDS, CONVERSION PROCESSES AND PRODUCTS

Feedstock	Conversion Process	Primary Energy Products	
Land-based biomass	Direct combustion		Thermal
Trees		Energy	Steam
Plants	Anaerobic digestion		Electric
Grasses			
			Combustibles
	Pyrolysis	Solid bio fuels	Chars
Water-based biomass			Methane
Single-cell algae	Gasification	Gaseous bio fuels	Hydrogen
Multicell algae			Low-Btu gas
Water plants	Direct liquefaction		Medium-Btu gas
			Light-hydro-carbon
			Hydrocarbon
Organic wastes	Ethanol production		Methanol
Municipal		Liquid bio fuels	Ethanol
Industrial			Hydrocarbons
Agricultural			Oils
Forestry			

(4)   gasification;

(5)   direct liquefaction;

(6)   ethanol production

Table 44 also exhibits the wide range of primary energy products from biomass. In general the energy products are either direct energy (steam and/or electricity) or "biofuels".

Economics can be analysed only in the scope of the specific combination of the three variables: the feedstock, the conversion process and the form of energy output. Moreover costs of biomass feedstocks vary significantly from region to region, such that any presentation of energy costs from biomass energy conversion technologies can only be imprecise. The technologies themselves are in highly varied states of commercial readiness: some fully mature, others still in conceptual stages. Even when demonstrations of newer technologies exist, it remains difficult to obtain precise cost figures because little data is available on closely monitored systems in commercial configurations. Finally, the diversity of sites, the varied economic relationships, and the economic value of trade-offs such as reduced waste disposal costs make any transnational comparisons of cost data somewhat hazardous.

Among the numerous combinations of these three variables, the following systems have already been economically operating in various IEA countries:

—   Small-scale systems (less than 100 kW) of land-based biomass combustion for space heating in the residential sector.

—   Combustion of wood and straw in medium-scale (less than 20 MW) mostly for district heating.

—   Combustion of industrial wastes in large-scale up to 250 MW in the forestry industry for the production of process heat, electricity and district heat.

—   Combustion of municipal waste in large scale for cogeneration of heat and electricty.

—   Anaerobic digestion of farm wastes including chicken litter for methane gas production in small scale.

In addition, gasification of various dry biomass for the production of process heat and fermentation of sugar and starch for alcohol production for motor fuel are often cited as technologies close to commercialization without government subsidies.

Contrary to other RETs, the fuel cost or the price of the feedstock accounts for a relatively large portion of the total production cost in many biomass systems. The cost of the feedstock varies widely reflecting the site-specific characteristics of biomass and relatively high transport costs. For instance, straw, one of the most popular biomass feedstocks, is by no means suitable for long-distance transport because it is too bulky compared with its energy content. Therefore, the cost of straw is quite sensitive to transport cost. On the other hand, costs of some other feedstocks like industrial and municipal wastes are null or even negative if the waste disposal cost could be avoided by the utilization of these wastes. This is especially the case in municipal waste applications.

Table 45 shows the costs of feedstocks for direct combustion assessed by Energy Technology Support Unit (ETSU) in the United Kingdom[1]. The cost varies from - £20/t of untreated refuse to £/52t of wood briquettes. ETSU also

TABLE 45

COSTS AND VALUES OF BIOFUEL FOR VARIOUS COMBUSTION SYSTEMS

(1981 price)

Biofuel	Cost of Supply (£/tonne)	*Gross Value (£/tonne)
Untreated Refuse	-20 - 0	5.60 - 10.50
Shredded/Screened Refuse	-17.50 - 6.50	13.80 - 22.00
Refuse Derived Fuel	-3.90 - 26.10	27.40 - 45.00
Industrial Waste	-20 - -10	25.20 - 41.50
Logs	0 - 35	
Industrial Use		11.00 - 25.70
Domestic Use		28.20 - 39.60
Wood Waste	-2 - 20	
Industrial Use		22.40 - 45.40
Domestic Use		45.50 - 63.90
Wood Briquettes	25 - 55	
Domestic Use		54.20 - 76.00
Baled Straw	7 - 22	
Industrial Use		20.50 - 45.60
Domestic Use		48.90 - 68.60
Straw Pellets	33 - 47	
Industrial Use		31.30 - 57.00
Domestic Use		54.20 - 76.00
Peat	0 - 40	
Industrial Use		22.70 - 36.70
Domestic Use		37.80 - 53.70
Chicken litter	-2 - 1	25.00 - 39.10

* The gross value represents the maximum price which could be charged for each solid biofuel taking into account the difference between the total system cost using coal and the non-fuel cost using biofuel.

Source: "Strategic Review of the Renewable Energy Technologies", Energy Technology Support Unit, London, November 1982.

1. "Strategic Review of the Renewable Energy Technologies - An Economic Assessment, Volume II", Energy Technology Support Unit, London, November 1982.

made an assessment of the gross value of various biofuels assuming that these fuels would replace coal in the heat market. The gross value represents the maximum price which could be charged for each solid biofuel taking into account the difference between the total system cost using coal and the non-fuel cost using biofuel. If the gross value exceeds the cost of biofuel, it is economic. According to the ETSU's assessment, the uses of refuse-derived fuels, industrial waste, wood waste and chicken litter are economic. In addition, baled straw and straw pellets are economic when they are used in domestic applications.

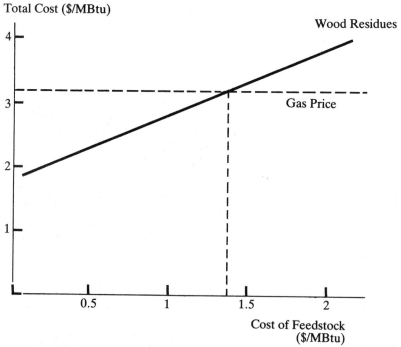

FIGURE 45
COST OF DIRECT COMBUSTION OF WOOD RESIDUES RETROFIT OF KILN FOR DRYING LUMBER IN CANADA

Assumptions:

System:	Heat supply system to a kiln with an annual lumber output of 59,000m³. Retrofit of an existing fossil-fuel-fired kiln
Annual Consumption of Sharings:	1976 Dt (37240 MBtu)
Capital Investment:	Can. $477210
O and M costs:	Can. $ 28370/year
Discount rate:	5%
Useful life:	20 years

Source: "ENFOR PROJECT C-258, a comparative assessment of Forest Biomass Conversion to Energy Forms", Energy, Mines and Resources, Canada, Ottawa, December 1983.

Many cases in line with the ETSU's study have been reported in other IEA countries. One of the most typical applications is found in the forestry-related industries such as paper and pulp and lumber. The case shown in Figure 45 is a system to produce heat for kiln drying of softwood lumber using wood residues in Canada. Since the conversion technology (combustion) is well established and the biggest uncertainty is the cost of feedstock, the figure shows the total heat cost by the cost of feedstock. The cost of delivered energy assuming a 95% system efficiency, 5% discount rate and 20-year system life is CDN $1.8/MBtu if feedstock is free. Since the natural gas price, which is replaced by this system, is CDN $3.2/MBtu, this system is economically viable even if feedstock is priced as high as CDN $1.4/MBtu.

Another example of a successful application of biomass is electricity production through the mass combustion of municipal waste. According to the data in Japan, electricity from combustion of municipal waste is estimated to cost 41 mills/kWh given a system useful life of twenty years and a 5% discount rate and can be cheaper than oil-fired power (Table 46).

TABLE 46
URBAN GARBAGE FUELLED POWER STATION IN JAPAN

Plant Size	600 t/day
Generating Capacity	5 000 kW
Capacity Factor	40%
Power Production p.a.	17.52 million kWh
Capital Investment**	1 845 million yen (US $7 687 500)*
O & M Cost	29.5 million yen p.a. (US $122 917)*
Generating Cost***	9.9 yen/kWh (US mills 41.43)*

(*) Exchange rate US $1 = 240 yen.
(**) Cost of boiler is assumed because it is already in use for garbage disposal.
(***) IEA Secretariat's estimate based on data provided by Japan for this review.

Source: Data provided by Japan for this review.

Anaerobic digestion of animal manure is widely utilized in many IEA countries. Table 47 shows the cost estimate of an anaerobic digester by ETSU in the United Kingdom. The estimate assumes an on-farm system utilizing pig and poultry waste. ETSU estimated that the system produced biogas at £1.9/GJ (£1.8/MBtu), while gas cost £2.4/GJ (£2.3/MBtu) in 1982 prices.

Besides these applications, gasification of various forms of dry biomass and fermentation of sugar and starch for alcohol production for motor fuel are promising; although the costs of these technologies are, in general, not yet

TABLE 47
COST OF ANAEROBIC DIGESTER IN UNITED KINGDOM
(1982 price)

Digester:	On-farm 200 m³
Feedstock:	Pig and Poultry Waste
	7% solids per tonne
Feedstock Input:	495 dry tonnes p.a.
Net Gas Output:	3.76 TJ p.a.
Annualised Cost* of	
Capital and O & M:	£7080 p.a.
Feedstock Cost:	Zero
Total Cost:	£1.9/GJ (£1.8/MBtu)
Price of Alternative Fuel:	Pipeline Gas £2.4/GJ

* Taking a 5% discount rate over a 20-year plant life.

Source: "Strategic Review of the Renewable Energy Technologies", Energy Technology Support Unit, London, November 1982.

competitive with conventional fuels. As shown in Table 48, according to the United States Office of Technology Assessment, the cost of alcohol produced from biomass exceeds the gasoline price in the United States by factors of 1.5 - 2.5 in 1980 prices. However, experts in the United States view that ethanol production based on fermentation of grains can achieve economic competitiveness without government subsidies as early as the late 1980s. In addition, the cost in the United States would be much lower if anhydrous ethanol were not required for the fuel blending. Clearly, methanol produced from wood is too expensive to compete with current technology, but lower feedstock costs and cheaper electric power costs plus a market for CO_2 produced could begin to improve the profitability of this technology.

Looking forward to energy plantation costs, analysis in the United States of potential energy crops (this example for Russian thistle) suitable for the south-west arid regions. Project costs (in US $1980) that range from

TABLE 48
COSTS OF BIOFUEL AS DELIVERED TO THE AUTO-SERVICE STATIONS IN THE UNITED STATES
(in 1980 US $)

Ethanol from corn	$12.50 - 17.60/MBtu
Methanol from wood	$13.40 - 22.00/MBtu
Methanol from herbage	$16.50 - 25.00/MBtu

Gasoline price in the U.S. in 1980 was approximately $10/MBtu.

Source: U.S. Office of Technology Assessment. Data provided by the United States for this review.

US $2.30 to US $4.40/GJ — 4.2/MBtu assuming an energy content of 13.6 MJ/kg — were obtained. Estimates for the costs of extracted liquids range between US $8 and US $13.5/MBtu. These costs indicate that although significant development of technical approaches and infrastructure will be necessary, the economic potential appears attractive.

This brief treatment of costs is insufficient to present the realities of economic competitiveness for biomass energy resources. They serve to indicate that for some technologies, competitiveness has been achieved, while for others, such competitiveness may come in the future, dependent on technological improvements as well as increases in conventional energy prices. Real costs can only be determined when conditions in existing markets for biomass, whether agriculture or forestry based, are studied in detail and trade-offs and net costs fully understood. In general, the recent decline of petroleum prices has made many biomass energy options unattractive from the standpoint of cost of delivered energy. Nonetheless, direct combustion is thriving economically in many niche markets, and its market should continue to expand. Anaerobic digestion in agricultural settings is also proving its economic viability, and where wood and residues are available locally at favourable spot prices, gasification systems appear marginally economic.

4. *Passive Solar*

Passive solar systems have demonstrated their competitiveness with conventional fuels by saving 30% and more of a building's space heating requirements in some climates, though their effectiveness varies from one site to another. Public housing in Canberra, Australia, is reported to have achieved an energy saving of A $200 per annum with an additional building cost of A $600[1].

In the United States, analysis of costs associated with recently constructed passive solar buildings has revealed that in both new residential and new commercial passive solar building costs normalized on a per square foot basis are generally well within the range of costs of buildings of the same type but of conventional design. Figure 46 compares the cost of eleven experimental commercial building projects in the United States to the range of costs for similar buildings throughout the United States. While these buildings on average area-weighted basis consume 45% less energy than comparable buildings of conventional design, the differences in construction costs are appreciable only in a few cases. The exceptions occur where passive components are designed as isolated components, such as Trombe walls, or when the quantity of materials used in construction is increased significantly as in high mass designs.

Table 49 shows the cost estimates of various passive solar measures based on the data in the UK in 1982[2]. Simple measures like minimised overshading and window redistribution to a south elevation are quite cost

1. Data provided by Australia for this example.
2. "Strategic Review of the Renewable Technologies - An Economic Assessment, Volume II" Energy Technology Support Unit, the United Kingdom Department of Energy, November 1982.

TABLE 49

COSTS AND PERFORMANCE OF PASSIVE SOLAR HEATING MEASURES IN THE UNITED KINGDOM

(in 1984 £)

	Old, Uninsulated House			New, Insulated House		
	Investment (£)	Energy Saving (£/MBtu/yr)	Cost* (£/MBtu)	Investment (£)	Energy Saving £/MBtu/yr	Cost* (£/MBtu)
Simple Measures						
Correct orientation Minimised overshading	not applicable			0	2.4	0
Windows redistributed to south elevation	not applicable			73	3.8	1.2
Advanced Measures						
South-facing window area increased to 20m²	570	6.8	5.3	570	4.4	8.2
Conservatory covering 10m² of external wall	824	5.5	9.5	836	4.7	11.3
18m² roof space collector and fan	594	7.5	5.0	594	5.5	6.9
20m² double-glazed Trombe wall	1042	8.1	8.2	1697	6.8	15.8

Cost of Alternative Fuels in 1984 (£/MBtu)**

High fuel oil	7.8
Gas	4.1
Coal	4.6
Electricity	16.9

* IEA Secretariat estimates of levelised capital cost with a 5% discount rate and 30-year useful life. Both the amounts of investments and energy saving are based on the data provided by the United States for this review. The original figures on investment were inflated by 1.212 in order to take into account the inflation between 1981 and 1984.

** Energy prices for households in the UK in 1984 shown in "Energy Prices and Taxes - Fourth Quarter 1984" IEA/OECD. Efficiency losses of 10% and 25% are assumed for electricity and other fuels respectively.

effective. Moreover, some of the more advanced passive solar measures are also economically viable in comparison with the cost of fuel oil (US $7.8/MBtu in 1984). Passive solar systems are even more economical if they are replacing heat produced from electricity. While isolated state-of-the art passive solar components, such as Trombe Walls, may not be able to compete with coal in most cases, the advanced technologies described in Annex I that are now being developed in several IEA countries have the potential to improve the economics of isolated passive solar components.

5. Active Solar

Solar water heaters for domestic or service water can economically compete with conventional energy systems as an auxiliary source of energy in some IEA Member countries which have both high solar insolation and high costs of the energy sources displaced such as electricity. However, where this is not the case, active low temperature solar systems are generally not competitive with fossil-fuelled conventional water heating or space conditioning systems without tax credits or other subsidies. These systems need to be improved by a factor of 2-4 in performance or cost over today's state of the art to displace fossil-fuelled conventional energy systems in the United States and many other IEA countries[1] [2].

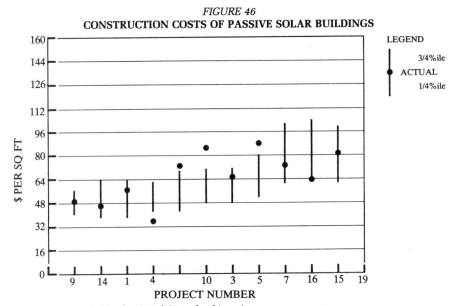

FIGURE 46
CONSTRUCTION COSTS OF PASSIVE SOLAR BUILDINGS

Source: Data provided by the United States for this review.

Figure 47 shows an IEA Secretariat estimate of the levelised cost of an active evacuated-tube solar water heating system by the amount of solar radiation based on the level of current active solar technologies. At sites with an average daily insolation of 5 kWh/m^3, the cost of the system is US \$18.0/ MBtu which is more than twice as high as the gas price of approximately US \$8/MBtu for United States households in 1984[2]. However, in a country like Japan where the gas price for households is much higher (approximately US \$21/MBtu) than the United States, such a system can be competitive with gas[2]. In terms of payback period, the United States case replacing gas in the residential sector implies approximately 26 years, while the Japanese case is equal to approximately eleven years.

1. "Strategic Review of the Renewable Energy Technologies", Energy Technology Support Unit, the UK DOE, November 1982.
2. "Energy Prices and Taxes - Fourth Quarter 1984" IEA/OECD, Paris, 1985.
3. "Energy Prices and Taxes - Fourth Quarter 1984" IEA/OECD, Paris, 1985.

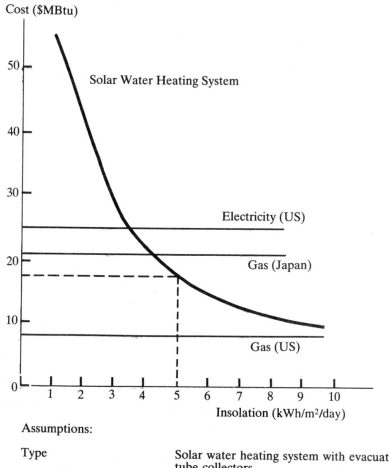

FIGURE 47
CURRENT COST OF ACTIVE SOLAR WATER
HEATING SYSTEM BY INSOLATION

Cost ($MBtu)

Solar Water Heating System

Electricity (US)

Gas (Japan)

Gas (US)

Insolation (kWh/m²/day)

Assumptions:

Type	Solar water heating system with evacuated tube collectors
Collector area	4m²
System efficiency	35%
System useful life	20 years
Capital investment	$500/m²
Discount rate	5%
Annual output	8.7 Mbtu at 5kWh/m²/day

Source: IEA calculations based on the data provided by the United States and Japan for this Review. No O&M costs assumed.

On the other hand, since electricity prices for households amounted to US $24-25/MBtu in the United States in 1984, the same system can be economically viable if the only alternative available is electricity.

As shown in Table 50, solar insolation varies considerably from one country to another with large seasonal variations of insolation in some countries especially in the United Kingdom. Due to these seasonal variations, active solar systems remain an auxiliary source of energy in almost all countries.

TABLE 50
INSOLATION IN VARIOUS COUNTRIES
$(kWh/m^2/day)$

	Midsummer (A)	Midwinter (B)	(A/B)	Annual Mean
UK	5.0	0.5	10.0	2.5
Central USA	7.2	3.1	2.4	5.3
South France	6.7	1.4	4.8	4.2
Israel	8.6	3.1	3.0	6.1
Australia	6.2	3.6	1.8	5.5
Japan	4.7	1.9	2.4	3.6

Source: "Strategic Review of the Renewable Energy Technologies", Energy Technology Support Unit, UK Department of Energy, 1982.

The capital investment for larger systems for large multi-family buildings or commercial or industry process heat projects can be lower than US $500/m^2$ assumed in the example of a single-family residential application. Larger systems would still need back-up systems or heat storage devices to cope with the seasonal variation in insolation. Although no detailed specification is available, data from United States Department of Energy (Figure 48) show that seasonal storage costs of energy could range from US $0.90 per MBtu for aquifer storage for US $11.70 per MBtu for steel tanks. Therefore, the economics of large systems is much more complicated than the simple case shown in Figure 48.

Table 51 shows that space heating and cooling are presently not economic. The table also shows the prospects of cost reductions in solar water-heating, water-space heating and cooling systems in the United States by the year 2000. The costs in the table were calculated by the IEA Secretariat based on the capital investment and annual production data provided by the United States DOE. Although no detailed specifications are available, substantial cost reductions are expected in all three applications. The expectation of cost reduction mainly comes from the decline in capital investment. A large portion of the reduction in capital investment is a reflection of mass production. According to the opinion of experts, the cost of collectors could be cut by half if the production increased fivefold. Compared to the prices of alternative energy sources (US $8/MBtu for gas, and US $24-25/MBtu for electricity in the United States in 1984), economic viability will still be limited mainly to the cases of water heating systems replacing electricity in this period.

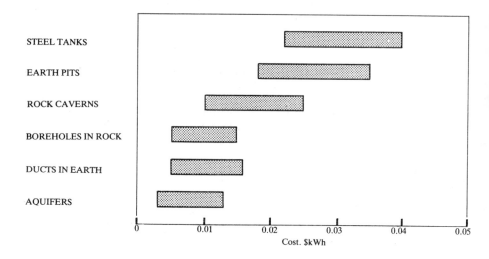

6.   *Tidal Energy*

Tidal energy technology has already demonstrated its economic viability in France. According to Electricité de France (EDF), its 240 MW tidal power station at La Rance costs almost the same as nuclear-fuelled generation. EDF estimates that a new tidal power system of the same size could compete with coal generation. According to "New Energy Overseas Information", Sept. 1984 by the New Energy Development Organization in Japan, generating costs of EDF are:

Tidal power at La Rance	12.5 centimes/kWh
New tidal power plant	31 centimes/kWh
Nuclear power	11.5 centimes/kWh
Hydro power	14 centimes/kWh
Coal-fired	30 centimes/kWh
Oil-fired	50 centimes/kWh

Besides the experience in France, feasibility studies at various sites including Canada and the United Kingdom have been carried out. For instance, a pre-feasibility study on the Severn Estuary tidal power project in the United Kingdom published in 1981 reported that the generating cost of the project could be 3.1p/kWh in December 1980 price, which was approximately 50% higher than the generating cost of a coal-fired power plant at that time and almost the same as that of an oil-fired plant[1].

---

1.   "Tidal Power from the Severn Estuary", The Severn Barrage Committee, London, 1981.

However, tidal systems would produce electricity for only several hours at times largely controlled by the tidal cycle which on many occasions would not necessarily contribute to electricity generation at a time of high demand. Thus, tidal systems as presently constructed do not provide firm capacity. In other words, the cost of tidal energy generation should be compared to only the fuel costs of conventional generation which were 1.2-1.5p/kWh for coal and 2.0-2.5p/kWh for oil in the fourth quarter of 1980. Using this approach, tidal generation would be twice as costly as coal fuel costs and 24% higher than oil generating fuel costs.

As for the future outlook, since tidal energy technology is relatively mature and costs are site-specific, careful site selection along with efforts for capital cost reduction through rigorous control of construction work at various stages is important. Figure 49 shows the effect of capital cost on total generating cost based on the Severn data. Given the 1984 fuel cost of a coal-fired plant in the United Kingdom, the capital investment of the project should be around £3 000 million or £420/kW in 1984 prices so that the project could become fully competitive with existing coal-fired plants.

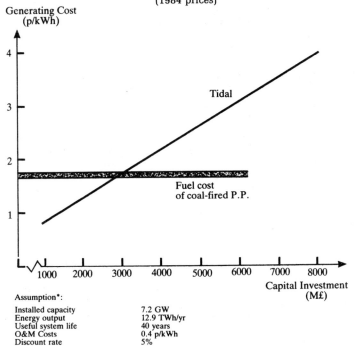

FIGURE 49
GENERATING COST OF TIDAL POWER SYSTEM BY CAPITAL INVESTMENT
(1984 prices)

Assumption*:

Installed capacity	7.2 GW
Energy output	12.9 TWh/yr
Useful system life	40 years
O&M Costs	0.4 p/kWh
Discount rate	5%

* The same as those assumed for the Severn inner barrage project except for O&M costs. O&M costs were assumed to be £35 million/year or 0.27 p/kWh in December 1980 prices in the Severn report.

Source: IEA calculations based on the data in "Tidal Power from the Severn Estuary", the Severn Barrage Committee.

7. Solar Thermal

The economic viability of active high temperature solar thermal technologies has not been demonstrated yet. The specific costs of present solar thermal systems are by a factor of two to six higher than conventional alternatives[1][2]. United States' studies claim approximately 130 mills/kWh for both the central receiver system and for the parabolic dish system based upon the present state of the art with a discount rate of 3.15% and certain favourable tax benefits[3]. The parabolic trough system can provide process heat of 100-315°C at US $27/MBtu, which is approximately two to four times as high as the cost of process heat produced by natural gas in the United States in 1984 [3].

In 1984, the Zürich Municipal Utility and a consortium of Swiss companies (SOTEL) completed a feasibility study for a central receiver plant in the Maroz Valley in Switzerland. The study concluded that with a more favourable site, a plant 50 MW$_e$ in size and 4 000 hours of operation under peak conditions per year, generating costs would be near US $0.16/kWh.

The 1986 United States Department of Energy Five Year Research and Development Plan for solar thermal technology estimates that current central receiver systems would have an installed cost of US $2 900/kW$_e$ (1984$) with an annual plant efficiency of 17% and the plant life of thirty years. As shown in Table 52, the generation cost of such a system at a 5% discount rate without any tax benefits is approximately US ¢22.6/kWh.

A salt gradient solar pond, 2 000 m^2 in area, for a thermal application in the United States was built for US $38/m^2. An estimate for the current energy cost of solar ponds for thermal applications at an average site, constructed at US $28/m^2, is US $5.9/MBtu[3].

For electricity production, very large ponds should be less expensive per unit area, costing in the vicinity of US $20 m^2. The Electric Power Research Institute in the United States concluded that a 50 MW$_e$ salt gradient pond would have a capital cost of US $5 400/kW$_e$ and generate electricity for US $0.26/kWh.

With regard to the future possibilities, the United States government foresees a further cost reduction of the parabolic trough system from US $27/MBtu to US $9/MBtu in 1984 price by 1990. A significant decline in the generating cost of solar thermal technologies is also projected by the United States government. As shown in Table 52, the capital investment of a 100 MW central receiver system could be reduced to US $2 200/kW (in 1984 price) by 1995 with improvements in efficiency, O & M costs and availability.

---

1. Data provided by the United States for this review.
2. Handbook of Energy Technology and Economics, Chapter 15, *Solar Thermal Energy Conversion* by Kenneth C. Brown.
3. Data provided by the United States for this review.

TABLE 51
FUTURE PROSPECTS OF ACTIVE SOLAR SYSTEMS IN THE UNITED STATES
(in 1983 US $)

		Current Status	Future Prospects
Application			
A. Alternative Energy - Electricity			
Water Heating	Capital Investment ($)	3 600	1 500
	Annual Savings (MBtu)	13	13
	Cost*($/MBtu)	21.7 (22.6)***	9.0 (9.4)***
	Payback Period**(years)	11.3	4.7
Water and Space Heating	Capital Investment ($)	12 500	5 000
	Annual Savings (MBtu)	26	23
	Cost* ($/MBtu)	37.7 (39.3)***	17.0 (17.7)***
	Payback Period**(years)	19.7	7.9
Cooling	Capital Investment ($)	168 000	35 000
	Annual Savings (MBtu)	70	57.3
	Cost* ($/MBtu)	188.0 (196.0)***	47.8 (49.8)***
	Payback Period**(years)	98.4	25.0
B. Alternative Energy - Gas			
Water Heating	Capital Investment ($)	3 600	1 500
	Annual Savings (MBtu)	17.4	14.4
	Cost* ($/MBtu)	16.2 (16.9)***	8.2 (8.6)***
	Payback Period**(years)	26.5	13.4
Water and Space Heating	Capital Investment ($)	12 500	5 000
	Annual Savings (M/Btu)	57	47
	Cost* ($/MBtu)	17.2 (17.9)***	8.4 (8.8)***
	Payback Period**(years)	28.1	13.6

*   IEA Secretariat estimates of levelised capital cost with a 5% discount rate and 20-year useful life. Costs of alternative energy sources: Electricity $24.4/MBtu, gas $7.8/MBtu.

**   IEA Secretariat estimate based on the assumptions of no O&M costs and $24.4/MBtu for electricity and $7.8/MBtu for gas as the costs of alternative energy systems.

*** Costs in 1984 price

Source: Data provided by the United States for this review.

Both mass production and the economies of scale play important roles for these cost reductions. The past experience in heliostat production costs exhibits a learning curve of about 86% in the United States[1]. That is, as the accumulated number of production doubles, the production cost declines by 14%. With respect to economies of scale, systems of 10 MW or larger are often considered the minimum size needed for standard utility applications. For instance, a study by an IEA sub-group shows that a 100 MW central receiver system could be as much as 75% cheaper than a 10 MW system[2].

1. Data provided by the United States for this review.
2. "A model for the economic assessment of solar power plants".
    Report initiated by the Executive Committee of the IEA SSPS project, Prof. Dr. G. Faninger, Stuttgart, September 1984.

TABLE 52
LONG-TERM (1995) TECHNICAL GOALS FOR CRS IN THE UNITED STATES
100MW PLANT
(in 1984 US$)

	Current Status[1]	1995 Goal
Capital Investment (M$)	290 ($2 900/kW)	220 ($2 200/kW)
Average Total Efficiency (%)	17	22
Availability (%)	90	95
Annual Production (GWh)	108	148
Generating Cost (¢/kWh)	22.61	11.5

Assumptions:

Rated power	100 MW
Solar direct isolation	2690 kWh/m² year (7.4 kWh/m²-day)
Heliostat	571 200 m²
Annual O&M costs	$6 M for the current and $3 M for the 1995 estimate
System useful life	30 years
Discount rate	5%

(1) The five-year plan shows that the current system would deliver energy over the life of the plant at a cost of US $0.13 per kWh with certain favourable tax benefits and a discount rate of 3.15% rather than 5% assumed for Table 52.

The above examples illustrate the diversity of approaches that are still being tried to demonstrate the economic feasibility of solar thermal applications. In any case, uncertainties as to the future projections are still high and, undoubtedly, further R&D activities are necessary to achieve these goals.

8.   Photovoltaics

Photovoltaics (PV) system costs are not yet competitive in most cases. At present PV systems are economic only for stand-alone power systems serving discrete energy demands such as consumer products, remote village power, navigational aids and remote microwave repeaters whose costs of alternatives are high and/or maintenance presents difficulties. According to reports from the United States Department of Energy, PV generating systems cost US $0.50/kWh in 1985 terms[1].

The capital cost of a photovoltaic system depends on the system power rating, the ease of transportation to, and installation at the site, the need for special hardware such as energy storage, and the size of the system.

---

1.   Data provided by the U.S. for this review.

Generation costs vary considerably depending on the amount of solar radiation. Figure 50 shows an estimate of the levelised generation costs of PV systems by the amount of solar radiation based on the level of the current PV technology. The PV generation costs US $0.50/kWh in the south-western part of the United States in which the amount of radiation is as high as 6 kWh/m²/day, while it costs more than US $0.80/kWh in the region with an insolation of less than 4 kWh/m²/day such as in the south-east United States (Figure 51).

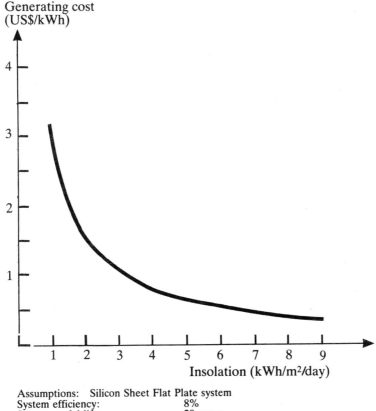

FIGURE 50
CURRENT PV GENERATING COST BY INSOLATION

Generating cost (US$/kWh)

Insolation (kWh/m²/day)

Assumptions: Silicon Sheet Flat Plate system
System efficiency:	8%
System useful life:	20 years
System size	10 peak kW
Capital investment:	$11145/kW
Operating cost:	$50/kW/year
Discount rate:	5%
Availability:	80%

Source: IEA calculations based on the data provided by the United States, Japan and Germany.

Unit costs for large utility-connected systems above 500 kW could be lower than those which are shown in Table 53. According to another estimate by United States Department of Energy, such large systems can be constructed

FIGURE 51
INSOLATION IN THE UNITED STATES

TOTAL HORIZONTAL RADIATION
ANNUAL AVERAGE-DAY VALUES
kWh/m²/day

Source: "Assessment of Distributed Photovoltaic Electric Power Systems", EPRI, October 1980.

at US $7000 - 8000/kW, which roughly corresponds to total generating costs of US $0.30 - 0.35/kWh. It should be noted that these estimates do not include the costs of battery storage. The generating costs of these photovoltaic systems could increase by almost 50% if the battery storage is added (see Annex I for a detailed explanation of the costs of the battery storage).

In any case, given the generation cost of new oil-fired power stations of US $0.05-0.06/kWh, the cost of PV systems is almost six to ten times as high as that of oil-fired power plants even in regions with high insolation.

With regard to the possibilities of future cost reductions, three factors are of importance: capital investment, system life expectancy and conversion efficiency. According to the plan of the United States Department of Energy shown in Table 53, module cost is planned to be reduced by factors of 7-14 for flat-plate systems and 5-8 for concentrator systems by the late 1990s. The investment for the balance of the system is also expected to decline by approximately 80% during the same period. As a result the total capital investment of PV systems per kW could be reduced from the present level of over US $10 000/kW to US $1 400 - 1 600/kW in the late 1990s. This reduction in capital investment will be achieved not only through technological breakthrough but also through mass production. Judging from the past experience, the production cost of the PV module could be reduced by 20%-30% if the accumulated production doubled.

System life expectancy, which is currently ten to twenty years, is also expected to increase to thirty years. If it were lengthened from twenty years to thirty years, the generation cost of PV systems could be reduced by nearly 20%.

System conversion efficiency is another factor which plays an important role for PV cost reduction. Under current technology, a system efficiency of 10% is achievable. If this were improved by 5 percentage points to 15%, the generation cost could be cut by one-third.

If all these goals shown in Table 53 were achieved, the generation cost of PV systems could be reduced from the present level of approximately US $0.50/kWh to about US $0.05/kWh in regions with solar radiation of

FIGURE 52
ECONOMIES OF SCALE OF OTEC SYSTEM

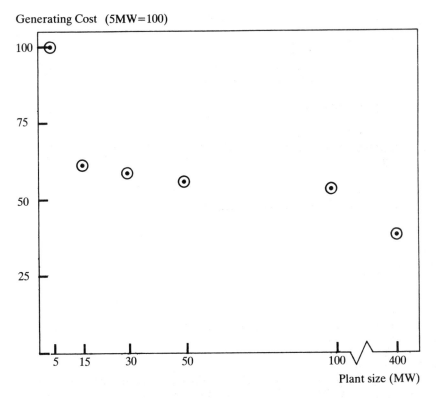

Generating Cost (5MW=100)

Plant size (MW)

Source: Data provided by the United States for this review.

6 kWh/m²/day such as the south-west United States by the second half of the 1990s. More than half of the total cost reduction is attributable to the decrease in capital investment, while the extension of useful life accounts for just over 10% of the reduction. The remainder comes from the improvement in efficiency. Thus it is possible that PV systems will penetrate into the field of central power station applications in the late 1990s at the earliest.

TABLE 53

**LONG-TERM (LATE 1990s) TECHNICAL GOALS FOR PHOTOVOLTAICS
IN THE UNITED STATES**

(in 1982 US$)

	Flat Plate Systems		Concentrator Systems	
	Current Status	Long-Term Goal	Current Status	Long-Term Goal
Capital Investment				
Module ($/m²)	550	40-75	750	90-100
Balance of System				
- area related ($/m²)	140	50	250	100
- power related ($/kW)	530	150	530	150
Total (1) ($/m²)	1 115	161-222	1 615	332-359
($/kW)	(11 150)	(1 379-1 451)	(11 212)	(1 602-1 374)
System Life Expectancy (years)				
	20	30	20	30
Module Efficiency (%)	12	13-17	17	23-29
System Efficiency (%)	10	11.7-15.3 (2)	14.4	20.7-26.1(2)
Generating Cost ($/kWh)(3)	0.450	0.046-0.049	0.452	0.054-0.046
	(0.484)[4]	(0.050-0.053)[4]	(0.487)[4]	(0.058-0.050)[4]

1) Including the indirect costs, which are assumed to be one-third of total capital investment.
2) A 10% loss is assumed between the module output (DC) and the system output (AC).
3) Solar insolation 6kWh/m²/day, discount rate 5%.
4) Costs in 1984 price

Source: Secretariat's estimates based on the data provided by the United States for this review.

9.   *Wave*

Although small scale devices for navigational buoys are economically feasible, wave energy technology has not progressed sufficiently to demonstrate its economic viability.

The second stage "Kaimei" in Japan is estimated to be about twice as costly as diesel engine generating systems in isolated locations[1]. In the United Kingdom, the National Engineering Laboratory is reported to expect that the maximum generating cost estimate of their wave energy breakwater system for the Scottish location (the Island of Lewis) is US ¢7.34/kWh[2].

In Norway, two firms, which are now operating wave energy systems, are expecting to produce power at a cost lower than the projected NKR 0.5-0.6/kWh (US $0.067-0.08/kWh, NKR 7.5 = US $1)[3].

In summary, estimates from feasibility studies indicate that wave energy technology could be competitive with diesel engine generating systems in isolated locations. However, the economic viability of these projects has not been actually demonstrated yet, and some other papers report much more conservative cost figures[4][5].

Further improvement in conversion efficiency together with the efforts for the reduction of capital investment will probably be required. In general, however, it is premature to predict economic performance for wave energy.

10.   OTEC

Although studies in France, Japan and the United States suggest that large scale OTEC could be competitive with conventional energy sources, generating costs at current OTEC experiment systems seem to exceed the cost of conventional energy sources by a factor of more than ten[6].

In light of the current stagnant conditions of OTEC R&D activities, it is very difficult to predict future possibilities. Obviously further improvement in conversion efficiency and reduction in capital investment are required. These could be achieved through a scale-up in size. Figure 52 illustrates the economies of scale in OTEC. The generating cost of a 400 MW OTEC plant could be 61% lower per unit than that of a 5 MW plant[7]. This implies that the development of OTEC should be carried out at a large scale to establish its economic viability.

---

1. Data provided by Japan for this review.
2. "Wave power prototype nears construction phase" by Maurice Buggott and Richard Morris, Power Engineering, February 1985.
3. Data provided by Norway for this review.
4. According to "Research and Development on Air Turbine Wave Power Generator Systems" by M. Miyazaki and S. Ishii presented at ECOR International Conference in October 1984, the first stage "Kaimei" cost 340 yen/kWh or 20 times as high as the generating cost of oil-fired power plant.
5. "Wave Energy - The Department of Energy's R&D Program 1974-1983" by ETSU in United Kingdom, published in March 1985 concluded that the overall economic prospects for wave energy looked poor when compared with other electricity-producing renewable energy technologies.
6. Data provided by United States and Japan for this review.
7. Data provided by United States and Japan for this review.

TABLE 54
CURRENT STATUS OF RET

RET	Future Technology		Under Development		Commercialisation	
	Component Design	Bench Scale	Pilot Plant	Demon-stration Plant	Commercial with subsidies	Economic without subsidies

Solar
Photovoltaic
Active Solar

Passive Solar

Solar Thermal

- remote
- cooling
- domestic hot water
- isolated components
- integral design
- electricity solar pond central receiver dish electric
- IPH*/ electricity trough collectors

Wind
Small/medium
Large

Geothermal

Hydrothermal
Binary Cycle
Hot Dry Rock
Geopressured
Magma
Direct Heat

- flashed steam/ dry steam

Ocean

Tidal
Wave
OTEC

Biomass

Direct Combustion
Biogas
Ethanol
Gasification
Liquefaction

* IPH = Industrial Process Heat.

— 330 —

III. **CONCLUSIONS**

Generally speaking, technology development has the following stages:

1) Future-Technology
   a) Basic R&D for each component;
   b) Bench scale models at laboratory levels to demonstrate the technical viability of the technology.

2) Under-Development
   a) Pilot plants to clarify technical problems;
   b) Demonstration plants to clarify problems for commercialization.

3) Commercialization:
   a) Commercial-With-Incentives;
   b) Economic without subsidies.

Table 54 illustrates the current status of RETs by these classifications. In general only those technologies beyond the final stage of the demonstration phase can be the subject of detailed economic analyses. In other words, the data of other technologies which have not reached that stage yet are still insufficient and may not fully reflect their future potential. Therefore descriptions on the economics of these technologies not yet in the demonstration phase should be used with care.

In any case, some RETs are already at the final stage of the demonstration phase or even at the commercialization phase. Hydrothermal electricity/ direct heat applications of geothermal, active low temperature solar water heaters, passive solar, direct combustion of biomass fuels, biogas and tidal power technologies have been already commercialized without subsidies in many parts of the IEA. Small/medium wind energy conversion systems and methanol production have been commercialized as well.

Although any generalisation on the cost of each RET is inevitably subject to a wide range of variance due to site-specific characteristics of renewable energy sources, a large number of economic applications of RETs do exist throughout the IEA.

It should also be noted that requirements for cost reduction of RETs are different from one technology to another. Table 55 provides a summary of other probable areas where cost reduction can be achieved.

The following examples illustrate the types of cost reductions needed. Since availability of renewable energy sources are quite site-specific, the field of resource assessments is equally important. For instance, although the mechanical parts for hydrothermal applications of geothermal energy electricity have been already fully developed, resource assessment of geothermal reservoirs will be needed to increase the number of sites available for development. Another important R&D area is the improvement in the system configurations. For example, OTEC and wave systems need

R&D progress in this area. In addition some technologies, such as photovoltaic, active solar, solar thermal and wind, need larger volumes of production to achieve further cost reductions. Some other technologies are subject to economies of scale. That is the case in tidal power, OTEC, geothermal, solar thermal and some biomass technologies.

In summary, many RETs can be moved from the threshold of establishing economic viability and to economic utilization if appropriate government policy and R&D tools are applied in accordance with the requirements of each RET and in conjunction with industry activities.

TABLE 55
MOST IMPORTANT AREAS FOR IMPROVEMENT OF ECONOMIC VIABILITY OF RET

		R & D		Mass Production	Scale-up
	Resource Assessment	Configuration*	Other**		
Solar					
Photovoltaics	X	X	X		
Active Solar			X	X	
Passive Solar				X	
Solar Thermal	X			X	X
Wind					
Small/medium	X	X		X	
Large	X	X			
Geothermal					
Hydrothermal					
Electricity	X				
Direct Heat	X				X
Binary Cycle	X				X
Hot Dry Rock	X		X		
Geopressured	X		X		
Magma	X		X		
Ocean					
Tidal	X				X
Wave	X	X	X		
OTEC	X	X	X		X
Biomass					
Direct Combustion					
Biogas		X			
Ethanol			X		X
Gasification			X		X
Liquefaction			X		X

* Including the matching between demand and supply.
** Including new technological breakthroughs.

APPENDIX

REVIEW ADVISORY BOARD
(RAB)

Chairman: Frederick H. Morse
US Department of Energy
Washington D.C. 20585
UNITED STATES

Jay Barclay
Renewable Energy Division
Dept. of Energy, Mines
and Resources
580 Booth Street
Ottawa K1A OE4
CANADA

T.S. Crawford
State Energy Commission
Energy Information Centre
Box L921-G.P.O.
Perth
AUSTRALIA

Bernd Dietrich
Bundesverband Solarenergie
Kruppstrasse 5
D-4300 Essen 1
GERMANY

Horst Hörster
Philips Research Laboratory
P.O. Box 1980
D-51 Aachen
GERMANY

Lars Astrand
UKAB
P.O/ Box 125
S-75104 Uppsala
SWEDEN

W. H. Bloss
University of Stuttgart
Pfaffenwaldring
D-700 Stuttgart 80
GERMANY

R. Clare
Sir Robert MacAlpine & Co
40 Bernard Street
London WC1 1LG
UNITED KINGDOM

Takuya Homma
University of Tsukuba
Sakura
Niihfari
Ibaraki, 305
JAPAN

Masakazu Kobayashi
New Energy Development Organisation
Sunshine 60 Bldg.
3-1-1, Higashi Ikebukuro
Toshima-ku
Tokyo 170
JAPAN

Carlo La Porta
Solar Energy Industries
  Association
Suite 503
1717 Massachusetts Ave N.W.
Washington DC 20036
UNITED STATES

Nigel C. McKenzie
Kleinworth, Benson Ltd
20 Fenchurch Street
London EC3
UNITED KINGDOM

Brasford Mead
The Baron Group
30 Tower LN
Avon CT 06001
UNITED STATES

Robert Mertens
Katholieke Universiteit Leuven
Kardinaal Mercier Laan 94
B-3030 Heverlee
BELGIUM

Johan Nijs
Katholieke Universiteit Leuven
Kardinaal Mercier Laan 94
B-3030 Heverlee
BELGIUM

Enrique Ochavan
Ministry of Industry and Energy
Paseo de la Castellana 160
E-28071 Madrid
SPAIN

Ignacio Oyarzabal
Ministry of Industry
  and Energy
Paseo de la Castellana 160
E-28071 Madrid
SPAIN

Keijo Sahrman
Technical Research Centre
  of Finland
P.O. Box 221
SF-40101 Jyväskylä 10
FINLAND

Bernd Stoy
Bundesverband Solarenergie
Kruppstrasse 5
D-4300 Essen 1
GERMANY

Shiego Suzuki
New Energy Development Organisation
Sunshine 60 Bldg.
3-1-1, Higashi Ikebukuro
Toshima-ku
Tokyo 170
JAPAN

The Members of the RAB contributed as experts in their field and did not necessarily represent the opinion of their government or business.

OECD SALES AGENTS
DÉPOSITAIRES DES PUBLICATIONS DE L'OCDE

ARGENTINA - ARGENTINE
Carlos Hirsch S.R.L.,
Florida 165, 4° Piso,
(Galeria Guemes) 1333 Buenos Aires
Tel. 33.1787.2391 y 30.7122

AUSTRALIA-AUSTRALIE
D.A. Book (Aust.) Pty. Ltd.
11-13 Station Street (P.O. Box 163)
Mitcham, Vic. 3132   Tel. (03) 873 4411

AUSTRIA - AUTRICHE
OECD Publications and Information Centre,
4 Simrockstrasse,
5300 Bonn (Germany)   Tel. (0228) 21.60.45
Local Agent:
Gerold & Co., Graben 31, Wien 1   Tel. 52.22.35

BELGIUM - BELGIQUE
Jean de Lannoy, Service Publications OCDE,
avenue du Roi 202
B-1060 Bruxelles   Tel. (02) 538.51.69

CANADA
Renouf Publishing Company Ltd/
Éditions Renouf Ltée,
1294 Algoma Road, Ottawa, Ont. K1B 3W8
Tel: (613) 741-4333
Toll Free/Sans Frais:
Ontario, Quebec, Maritimes:
1-800-267-1805
Western Canada, Newfoundland:
1-800-267-1826
Stores/Magasins:
61 rue Sparks St., Ottawa, Ont. K1P 5A6
Tel: (613) 238-8985
211 rue Yonge St., Toronto, Ont. M5B 1M4
Tel: (416) 363-3171
Sales Office/Bureau des Ventes:
7575 Trans Canada Hwy, Suite 305,
St. Laurent, Quebec H4T 1V6
Tel: (514) 335-9274

DENMARK - DANEMARK
Munksgaard Export and Subscription Service
35, Nørre Søgade, DK-1370 København K
Tel. +45.1.12.85.70

FINLAND - FINLANDE
Akateeminen Kirjakauppa,
Keskuskatu 1, 00100 Helsinki 10   Tel. 0.12141

FRANCE
OCDE/OECD
Mail Orders/Commandes par correspondance :
2, rue André-Pascal,
75775 Paris Cedex 16
Tel. (1) 45.24.82.00
Bookshop/Librairie : 33, rue Octave-Feuillet
75016 Paris
Tel. (1) 45.24.81.67 or/ou (1) 45.24.81.81
Principal correspondant :
Librairie de l'Université,
12a, rue Nazareth,
13602 Aix-en-Provence   Tel. 42.26.18.08

GERMANY - ALLEMAGNE
OECD Publications and Information Centre,
4 Simrockstrasse,
5300 Bonn   Tel. (0228) 21.60.45

GREECE - GRÈCE
Librairie Kauffmann,
28, rue du Stade, 105 64 Athens   Tel. 322.21.60

HONG KONG
Government Information Services,
Publications (Sales) Office,
Beaconsfield House, 4/F.,
Queen's Road Central

ICELAND - ISLANDE
Snæbjörn Jónsson & Co., h.f.,
Hafnarstræti 4 & 9,
P.O.B. 1131 – Reykjavik
Tel. 13133/14281/11936

INDIA - INDE
Oxford Book and Stationery Co.,
Scindia House, New Delhi 1   Tel. 45896
17 Park St., Calcutta 700016   Tel. 240832

INDONESIA - INDONÉSIE
Pdii-Lipi, P.O. Box 3065/JKT.Jakarta
Tel. 583467

IRELAND - IRLANDE
TDC Publishers - Library Suppliers,
12 North Frederick Street, Dublin 1.
Tel. 744835-749677

ITALY - ITALIE
Libreria Commissionaria Sansoni,
Via Lamarmora 45, 50121 Firenze
Tel. 579751/584468
Via Bartolini 29, 20155 Milano   Tel. 365083
Sub-depositari :
Editrice e Libreria Herder,
Piazza Montecitorio 120, 00186 Roma
Tel. 6794628
Libreria Hœpli,
Via Hœpli 5, 20121 Milano   Tel. 865446
Libreria Scientifica
Dott. Lucio de Biasio "Aeiou"
Via Meravigli 16, 20123 Milano   Tel. 807679
Libreria Lattes,
Via Garibaldi 3, 10122 Torino   Tel. 519274
La diffusione delle edizioni OCSE è inoltre
assicurata dalle migliori librerie nelle città più
importanti.

JAPAN - JAPON
OECD Publications and Information Centre,
Landic Akasaka Bldg., 2-3-4 Akasaka,
Minato-ku, Tokyo 107   Tel. 586.2016

KOREA - CORÉE
Kyobo Book Centre Co. Ltd.
P.O.Box: Kwang Hwa Moon 1658,
Seoul   Tel. (REP) 730.78.91

LEBANON - LIBAN
Documenta Scientifica/Redico,
Edison Building, Bliss St.,
P.O.B. 5641, Beirut   Tel. 354429-344425

MALAYSIA - MALAISIE
University of Malaya Co-operative Bookshop
Ltd.,
P.O.Box 1127, Jalan Pantai Baru,
Kuala Lumpur   Tel. 577701/577072

NETHERLANDS - PAYS-BAS
Staatsuitgeverij
Chr. Plantijnstraat, 2 Postbus 20014
2500 EA S-Gravenhage   Tel. 070-789911
Voor bestellingen: Tel. 070-789880

NEW ZEALAND - NOUVELLE-ZÉLANDE
Government Printing Office Bookshops:
Auckland: Retail Bookshop, 25 Rutland Street,
Mail Orders, 85 Beach Road
Private Bag C.P.O.
Hamilton: Retail: Ward Street,
Mail Orders, P.O. Box 857
Wellington: Retail, Mulgrave Street, (Head
Office)
Cubacade World Trade Centre,
Mail Orders, Private Bag
Christchurch: Retail, 159 Hereford Street,
Mail Orders, Private Bag
Dunedin: Retail, Princes Street,
Mail Orders, P.O. Box 1104

NORWAY - NORVÈGE
Tanum-Karl Johan
Karl Johans gate 43, Oslo 1
PB 1177 Sentrum, 0107 Oslo 1Tel. (02) 42.93.10

PAKISTAN
Mirza Book Agency
65 Shahrah Quaid-E-Azam, Lahore 3 Tel. 66839

PORTUGAL
Livraria Portugal,
Rua do Carmo 70-74, 1117 Lisboa Codex.
Tel. 360582/3

SINGAPORE - SINGAPOUR
Information Publications Pte Ltd
Pei-Fu Industrial Building,
24 New Industrial Road No. 02-06
Singapore 1953   Tel. 2831786, 2831798

SPAIN - ESPAGNE
Mundi-Prensa Libros, S.A.,
Castelló 37, Apartado 1223, Madrid-28001
Tel. 431.33.99
Libreria Bosch, Ronda Universidad 11,
Barcelona 7   Tel. 317.53.08/317.53.58

SWEDEN - SUÈDE
AB CE Fritzes Kungl. Hovbokhandel,
Box 16356, S 103 27 STH,
Regeringsgatan 12,
DS Stockholm   Tel. (08) 23.89.00
Subscription Agency/Abonnements:
Wennergren-Williams AB,
Box 30004, S104 25 Stockholm.
Tel. (08)54.12.00

SWITZERLAND - SUISSE
OECD Publications and Information Centre,
4 Simrockstrasse,
5300 Bonn (Germany)   Tel. (0228) 21.60.45
Local Agent:
Librairie Payot,
6 rue Grenus, 1211 Genève 11
Tel. (022) 31.89.50

TAIWAN - FORMOSE
Good Faith Worldwide Int'l Co., Ltd.
9th floor, No. 118, Sec.2
Chung Hsiao E. Road
Taipei   Tel. 391.7396/391.7397

THAILAND - THAILANDE
Suksit Siam Co., Ltd.,
1715 Rama IV Rd.,
Samyam Bangkok 5   Tel. 2511630

TURKEY - TURQUIE
Kültur Yayinlari Is-Türk Ltd. Sti.
Atatürk Bulvari No: 191/Kat. 21
Kavaklidere/Ankara   Tel. 25.07.60
Dolmabahce Cad. No: 29
Besiktas/Istanbul   Tel. 160.71.88

UNITED KINGDOM - ROYAUME-UNI
H.M. Stationery Office,
Postal orders only:
P.O.B. 276, London SW8 5DT
Telephone orders: (01) 622.3316, or
Personal callers:
49 High Holborn, London WC1V 6HB
Branches at: Belfast, Birmingham,
Bristol, Edinburgh, Manchester

UNITED STATES - ÉTATS-UNIS
OECD Publications and Information Centre,
2001 L Street, N.W., Suite 700,
Washington, D.C. 20036 - 4095
Tel. (202) 785.6323

VENEZUELA
Libreria del Este,
Avda F. Miranda 52, Aptdo. 60337,
Edificio Galipan, Caracas 106
Tel. 32.23.01/33.26.04/31.58.38

YUGOSLAVIA - YOUGOSLAVIE
Jugoslovenska Knjiga, Knez Mihajlova 2,
P.O.B. 36, Beograd   Tel. 621.992

Orders and inquiries from countries where Sales
Agents have not yet been appointed should be sent
to:
OECD, Publications Service, Sales and
Distribution Division, 2, rue André-Pascal, 75775
PARIS CEDEX 16.

Les commandes provenant de pays où l'OCDE n'a
pas encore désigné de dépositaire peuvent être
adressées à :
OCDE, Service des Publications. Division des
Ventes et Distribution. 2. rue André-Pascal. 75775
PARIS CEDEX 16.

70595-03-1987

OECD PUBLICATIONS, 2, rue André-Pascal, 75775 PARIS CEDEX 16 - No. 43851 1987
PRINTED IN FRANCE
(61 87 06 1) ISBN 92-64-12942-1